SpringerBriefs in Earth System Sciences

Series Editors

Gerrit Lohmann
Jorge Rabassa
Justus Notholt
Lawrence A. Mysak
Vikram Unnithan

For further volumes:
http://www.springer.com/series/10032

Wolfgang Hiller · Reinhard Budich
René Redler

Earth System Modelling – Volume 6

ESM Data Archives in the Times of the Grid

 Springer

Wolfgang Hiller
Alfred-Wegener-Insitut für Polar- und
    Meeresforschung
Bremerhaven
Germany

Reinhard Budich
René Redler
Max-Planck-Institut für Meteorologie
Hamburg
Germany

ISSN 2191-589X   ISSN 2191-5903   (electronic)
ISBN 978-3-642-37243-8   ISBN 978-3-642-37244-5   (eBook)
DOI 10.1007/978-3-642-37244-5
Springer Heidelberg New York Dordrecht London

Library of Congress Control Number: 2011938123

Printed on acid-free paper

Springer is part of Springer Science+Business Media (www.springer.com)

Preface

Climate modelling in former times mostly covered the physical processes in the Earth's atmosphere. Nowadays, there is a general agreement that not only physical, but also chemical, biological and, in the near future, economical and sociological—the so-called anthropogenic—processes have to be taken into account on the way towards comprehensive Earth system models. Furthermore, these models include the oceans, the land surfaces and, so far to a lesser extent, the Earth's mantle. Between all these components feedback processes have to be described and simulated.

Today, a hierarchy of models exists for Earth system modelling. The spectrum reaches from conceptual models—back of the envelope calculations—over box-, process- or column-models, further to Earth system models of intermediate complexity and finally to comprehensive global circulation models of high resolution in space and time. Since the underlying mathematical equations in most cases do not have an analytical solution, they have to be solved numerically. This is only possible by applying sophisticated software tools, which increase in complexity from the simple to the more comprehensive models.

With this series of briefs on "Earth System Modelling" at hand we focus on Earth system models of high complexity. These models need to be designed, assembled, executed, evaluated and described, both in the processes they depict as well as in the results the experiments carried out with them produce. These models are conceptually assembled in a hierarchy of submodels, where process models are linked together to form one component of the Earth system (Atmosphere, Ocean, ...), and these components are then coupled together to Earth system models in different levels of completeness. The software packages of many process models comprise a few to many thousand lines of code, which results in a high complexity of the task to develop, optimise, maintain and apply these packages, when assembled to more or less complete Earth system models.

Running these models is an expensive business. Due to their complexity and the requirements with respect to the ratios of resolution versus extent in time and space, most of these models can only be executed on high performance computers, commonly called supercomputers. Even on today's supercomputers, typical model experiments take months to conclude. This makes it highly attractive to increase the efficiency of the codes. On the other hand, the lifetime of the codes exceeds the typical lifetime of computing systems and architectures roughly by a factor of 3. This means that the codes need not only be portable, but also constantly adapted to emerging computing technology. While in former times computing power of single processors—and that of clustered computers—was resulting mainly from increasing clock speeds of the CPUs, todays increases are only exploitable when the application programmer can make best use of the increasing parallelism off-core, on-core and in threads per core. This adds complexity to areas like IO performance, communication between cores or load balancing to the assignment at hand.

All these requirements put high demands on the programmers to apply software development techniques to the code, making it readable, flexible, well-structured, portable and reusable, but most of all capable in terms of performance. Fortunately, these requirements match very well an observation from many research centres: due to the typical structure of the staff of the research centres, code development oftentimes has to be done by scientific experts, who typically are not computing or software development experts. Therefore, the code they deliver needs a certain amount of quality control to assure fulfilment of the requirements mentioned above. This quality assurance has to be carried out by staff with profound knowledge and experience in scientific software development and a mixed background from computing and science.

Since such experts are rare, an approach to ensure high code quality is the introduction of common software infrastructures or frameworks. These entities attempt to deal with the problem by providing certain standards in terms of coding and interfaces, data formats and source management structures, that enable the code developers as much as the experimenters to deal with their Earth system models in a well acquainted, efficient way. The frameworks foster the exchange of codes between research institutions, the model inter-comparison projects so valuable for model development, and the flexibility of the scientists when moving from one institution to another, which is commonplace behaviour these days.

With an increasing awareness about the complexity of these various aspects, scientific programming has emerged as a rather new discipline in the field of Earth system modelling. At the same time, new journals are launched providing platforms to exchange new ideas and concepts in this field. Up to now we are not aware of any text book addressing this field, tailored to the specific problems the researcher is confronted with. To start a first initiative in this direction, we have compiled a series of six volumes, each dedicated to a specific topic the researcher is confronted with when approaching Earth System Modelling:

Volume 1: Recent Developments and Projects
Volume 2: Algorithms, Code Infrastructure and Optimisation
Volume 3: Coupling Software and Strategies
Volume 4: IO and Postprocessing
Volume 5: Tools for Configuring, Building and Running Models
Volume 6: ESM Data Archives in the Times of the Grid

This series aims at bridging the gap between IT solutions and Earth system science. The topics covered provide insight into state-of-the-art software solutions and in particular address coupling software and strategies in regional and global models, coupling infrastructure and data management, strategies and tools for pre- and post-processing and techniques to improve the model performance. So the series aims not only at the Earth system researcher already familiar with some modelling aspects, but also at all those people trying to get a comprehensive overview about some intrinsics of this field of science, on all levels of their career.

Volume 1 at hand familiarises the reader with the general frameworks and different approaches for assembling Earth system models. Volume 2 highlights major aspects of design issues that are related to the software development, its maintenance and performance. Volume 3 describes different technical attempts from the software point of view to solve the coupled problem. Once the coupled model is running, data are produced and postprocessed (Volume 4). The whole process of creating the software, running the model and processing the output is assembled into a workflow (Volume 5). Finally, this huge amount of data has to be archived and retrieved for later analysis. Volume 6 at hand describes coordinated approaches to help solve these complex tasks by employing Grid software and metadata standards.

Hamburg, February 2013   Reinhard Budich
                                                                         René Redler

Contents

Contributors

Gavin Bell Program for Climate Model Diagnosis and Intercomparison, Lawrence Livermore National Laboratory, Livermore, CA, USA, e-mail: bell51@llnl.gov

Benny Bräuer Alfred-Wegener-Institut für Polar- und Meeresforschung, Bremerhaven, Germany, e-mail: Benny.Braeuer@awi.de

Reinhard Budich Max Planck Institute for Meteorology, Hamburg, Germany, e-mail: reinhard.budich@mpimet.mpg.de

Luca Cinquini NASA Jet Propulsion Laboratory, Pasadena, CA, USA, e-mail: luca.cinquini@jpl.nasa.gov

Michael Diepenbroek MARUM-Zentrum für Marine Umweltwissenschaften, Bremen, Germany, e-mail: mdiepenbroek@pangaea.de

Peter Fox Rensselaer Polytechnic Institute, Troy, NY, USA, e-mail: pfox@cs.rpi.edu

Bernadette Fritzsch Alfred-Wegener-Institut für Polar- und Meeresforschung, Bremerhaven, Germany, e-mail: Bernadette.Fritzsch@awi.de

Robin Goldstone Lawrence Livermore National Laboratory, Livermore, CA, USA, e-mail: goldstone1@llnl.gov

John Harney Oak Ridge National Laboratory, Oak Ridge, TN, USA, e-mail: harneyjf@ornl.gov

Wolfgang Hiller Alfred-Wegener-Institut für Polar- und Meeresforschung, Bremerhaven, Germany, e-mail: Wolfgang.Hiller@awi.de

Stephan Kindermann Deutsches Klimarechenzentrum, Hamburg, Germany, e-mail: kindermann@dkrz.de

Michael Lautenschlager Deutsches Klimarechenzentrum, Hamburg, Germany, e-mail: lautenschlager@dkrz.de

Uwe Schindler MARUM-Zentrum für Marine Umweltwissenschaften, Bremen, Germany, e-mail: uschindler@pangaea.de

Martina Stockhause Deutsches Klimarechenzentrum, Hamburg, Germany, e-mail: stockhause@dkrz.de

Dean N. Williams Program for Climate Model Diagnosis and Intercomparison, Lawrence Livermore National Laboratory, Livermore, CA, USA, e-mail: williams13@llnl.gov

Chapter 1
ESM Data Archives in Times of the Grid

Reinhard Budich and Wolfgang Hiller

In this last book in a series of briefs on "Earth System Modelling" we are concerned with the treatment of data and metadata that are produced during the complete workflow of Earth system modeling. It becomes more and more evident that Earth system modeling faces a "data problem", resulting from at least 3 different observations:

- Massive increase in computing power available, and data produced;
- The ever increasing necessity to compare these data against all kinds of other data, being available locally or, in a growing number of cases, remotely; and
- The necessity to provide the data produced to a massively growing community of researchers and lay-men concerned with the impacts of climate change.

Whereas the first two aspects of this problem are also observed in other disciplines and often called the "Fourth Paradigm" or the "data tsunami" (see Hey et al. 2009), the problem is especially severe in Earth system modeling due to the third aspect: The requirements from the impacts scene mentioned above.

Other aspects arise within the intense modeling activities of the Intergovernmental Panel on Climate Change (IPCC) where different models and observational data are analyzed. Especially with the work for the fourth assessment report (AR4) in 2007 and recent work for AR5 it has been demonstrated, that an ensemble of different global coupled Earth system models (ESM) and thorough comparison and analysis of the model results with observational data from different sources, is a key ingredient in the identification of climate change. The tedious process of processing large amounts of high volume model data into the right data formats and under different metadata conventions has put a challenge on the agenda for the modeling centers worldwide,

R. Budich (✉)
Max-Planck-Institut für Meteorologie, Bundesstraße. 53, 20146 Hamburg, Germany
e-mail: reinhard.budich@mpimet.mpg.de

W. Hiller
Alfred-Wegener-Institut für Polar- und Meeresforschung, Am Handelshafen 12,
27570 Bremerhaven, Germany
e-mail: wolfgang.hiller@awi.de

W. Hiller et al., *Earth System Modelling – Volume 6*, SpringerBriefs in Earth System Sciences, DOI: 10.1007/978-3-642-37244-5_1, © The Author(s) 2013

namely to advance this process by means of distributed processing capabilities and uniform access mechanisms to ESM data. At the same time strict quality control has to be maintained, to ensure overall data integrity and accuracy.

Grid software has become an important enabling technology for several national and international climate community Grids worldwide. This brief is devoted to document these joint efforts, which have led to a new dimension of distributed data access and pre- and post-processing capabilities; it discusses these challenges as well as some of the solutions available to the data problem in Earth system modeling.

In Chap. 2 the contributing authors start with an overview of the worldwide scene of distributed archives, databases and data portals, which is quite hard, since it is very volatile in terms of its development. It shows that the Coupled Model Intercomparison Project Phase 5 (CMIP5) activities (further detail is given in Chap. 7) have a decisive, regulating influence on the activities of the community also in terms of their data management solutions. The chapter states some important postulations in terms of availability of the data, the catalogues about them, the processing power for processing the data, and some governance necessary to improve not only accessibility, but also security of data.

Then from Chap. 3 onward the focus is put on different technological aspects, starting with the harvesting of metadata with open access tools followed by a technology overview in the next chapter of data discovery architectures and techniques realized in times of the Grid. It shows what it takes a user today to identify, search and find specific data in the worldwide ESM data storage landscape - and how important standards are here. It also shows that a better worldwide coordination and governance in this respect would be extremely helpful.

Chapter 3 describes, at the example of the German C3Grid, a harvesting approach to establish a centralized metadata catalogue for different data collections with different underlying catalogues and metadata systems. Such approaches are prerequisite for all attempts to federate data collections. These data federations provide the necessary functionality and computing performance users need for their scientific work. They apply grid technology to build up a solid basis for subsequent workflows, which depend heavily not only on computing ressources, but also to a high extent on the correctness of data descriptions and metadata of these big data collections.

In Chap. 4 the user perspective is taken one step further in showing how, after finding the data, users can now access them. This step needs to consider the formats the data are available in, since many if not most users - especially when we consider "downstream" users from the impact side - are not ready to deal with data in GRIB or NetCDF formats. So a conversion of the "raw data" has to be organized, and server side processing or Grid based scheduled services are potential technologies to offer these conversions.

We then move over to different user driven data access mechanisms in Chap. 5, where also aspects of different local implementations and technical solutions for data access are discussed. While we have discussed until now relatively homogeneous data sources, with data originating from coupled runs of global Earth system model components, an improved understanding of climate change on the regional scale is also a central and indispensable element of current climate research, which immediately leads to more heterogeneous data scenarios. Therefore the concluding chapters give an overview of two existing national climate community Grids: In Chap. 6 the C3Grid from Germany is explained, where data selection as well as pre- and post-processing for heterogeneous data from different data sources together with subsequent workflow processing play a major role. Finally, in Chap. 7 the Earth System Grid from the United States of America is described: Here the focus is on distributed and uniform access to ESM data on the global scale.

An interesting new technology is described in more detail in Chap. 6 of this brief. The specialty of this solution is scheduled distributed processing, which is not (yet) offered in any other solution known to the authors so far. This results from the insight that networks today tend to be optimized for transaction processing and not for scientific workflows: Whereas "the Web" assumes rather short life times of connections and burst-wise transfer of rather small data entities, scientific workflows need to transfer rather large data entities, necessarily—given the low bandwidth available between sites—over long periods of time. So: Scientists rather avoid transferring larger amounts of data, and try to do their processing near to the data. A Grid based approach is a rather natural solution to this problem, as the chapter shows.

The next to last chapter of this brief deals with the Earth System Grid Federation. This is by all means the most influential, structure shaping activity in Earth system model data management. Not only does this CMIP5-activity make ESM data comparable via imposing the Climate Model Output Rewriter (CMOR) standard as quasi-standard for model inter-comparison projects, it also provides a tool for the integration of data nodes into a worldwide network of such nodes, and the portals which are central to the discover and access workflows of the scientists.

Chapter 8 concludes this brief. It shortly touches upon topics which the authors think will gain more importance in the future of "the grid", like network bandwidth and security issues, metadata, onthologies, workflows and Climate Knowledge Discovery. It does not touch the topic of networking scientists, which the authors recon to be a substantial part for a well-oiled grid. This would go far beyond the scope of this brief, but is partly tackled in Volume 1 of this series.

Reference

Hey T, Tansley S, Tolle KM (eds) (2009) The fourth paradigm: data-intensive scientific discovery. Microsoft Research

Chapter 2
Distributed Archives, Databases and Data Portals: The Scene

Michael Lautenschlager

In 2003 the online journal Computerwoche[1] referred a study of the University of California in Berkley about the increase of globally archived data: Every 3 years the amount of archived data doubles. The scientists counted two to three Exabyte in 1999 and five Exabyte in 2002. Data volumes in specific disciplines are much smaller than in others but the increase of annual production rates is comparable. In connection with the Fourth Assessment Report (AR4) of the Intergovernmental Panel on Climate Change (IPCC) Program for Climate Model Diagnostic and Intercomparison (PCMDI) at the Lawrence Livermore National Laboratory (LLNI) collected around 60 Terabyte of global climate model data. Data management for the next IPCC Assessment Report (AR5) which is scheduled for 2013 plans for at least 2 PetaByte (PB) of data related to the AR5.

The global climate model data base for AR5 is created within the Climate Modelling Intercomparison Project No. 5 (CMIP5). Data are produced by 30 international modelling centres which are running 60 global climate modelling configurations in total. 10 PB of climate global climate data are expected to be produced under CMIP5, 2 PB are requested from to modelling groups to be integrated in the international CMIP5 data federation and 1 PB of core data will be replicated at least between three data nodes (PCMDI (U.S.), BADC (Great Britain) and DKRZ/WDCC (Germany)) for direct usage in the IPCC process. But not only the total amount of climate model data is challenging, it is also the number of individual data entities. The CMIP5 data federation is projected to manage more than 5 Mio individual data files (in NetCDF/CF format) ranging in size from a few MegaByte to some GigaByte.

These data volumes infer an increase factor of 35 within 5 years. The data increase factor in Earth system modelling is considerably higher than for the general global data increase. This large increase imposes new strategies in data management and

---

[1] Computerwoche Online, 2003: Anzahl der gespeicherte Daten verdoppelte sich in drei Jahren. Meldung 31.10.2003, 13:47.

M. Lautenschlager (✉)
Deutsches Klimarechenzentrum, Hamburg, Germany
e-mail: lautenschlager@dkrz.de

W. Hiller et al., *Earth System Modelling – Volume 6*, SpringerBriefs in Earth System Sciences, DOI: 10.1007/978-3-642-37244-5_2, © The Author(s) 2013

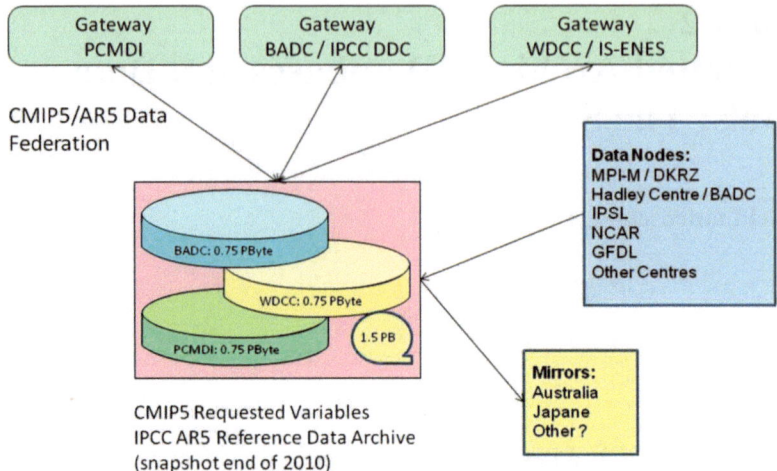

Fig. 2.1 CMIP5/AR5 data management concept. The data infrastructure for the next IPCC assessment report is planned as a federation of three geographically distributed data archives: PCMDI (Program for Climate Model Diagnosis and Intercomparison at LLNL, US), BADC (British Atmospheric Data Centre, UK) and WDCC (World Data Center Climate at DKRZ, G). Transparent data access will be established by web-portals (Gateways), more data will be provided to the federation by additional data nodes

storage. For the IPCC AR5 data management the central data archive is replaced by a global data federation with three core archives for replication of core data (Fig. 2.1).

Currently available Web-technologies (Web 2.0) create expectance of transparent data and information access. The Organisation for Economic Co-operation and Development (OECD) sets the scene[2]: "…Innovative scientific research has a crucial role in addressing global challenges—ranging from health care and climate change to renewable energy and natural resources management. The speed and depth of this research depends on fostering collaborative exchanges between different communities and assuring its widest dissemination. The exchange of ideas, knowledge and data emerging is fundamental for human progress and is part of the core of OECD values. Thus, I am very pleased that the OECD has taken the lead in developing principles and standards to facilitate access to research data generated with public funding. The rapid development in computing technology and the Internet have opened up new applications for the basic sources of research—the base material of research data—which has given a major impetus to scientific work in recent years. Databases are rapidly becoming an essential part of the infrastructure of the global science system …". Scientific primary data are raw material to gain new knowledge but the realization of the OECD data access picture imposes the change from domain specific to interdisciplinary data access and information exchange.

Modern data infrastructures (computer, disks and networks) together with state of the art software infrastructures (Web 2.0 and Grid-technology) allow in principle for

---

[2] http://www.oecd.org/dataoecd/9/61/38500813.pdf

transparent access to large amounts of data. The resulting interdisciplinary data usage imposes data management constraints which include legal aspects like copyright, access constraints and person information protection.

A number of key functionalities are required by data users and data providers:

Data archives must be available in the World Wide Web. Transparent and seamless access to quality proven data is expected. Data archives with Web-based data access have been developed for the last 15–20 years (Lautenschlager and Reinke 1997). The development started with community specific data systems to support scientific work and long-term archiving within the climate community. Long-term archiving and quality assurance together with the related data curation aspects imposed even 15 years ago sharing of responsibilities between topic related data archives. Two basic data types are distinguished in Earth system research, data from individual instrumental observations and data from numerical models, satellites and monitoring systems. Each data type has specific characteristics and data management requirements. Individual observational data are less in volume but heterogeneous in structure which imposes more complexity in metadata handling and quality assurance. Data from models, satellites and monitoring are large in volume but homogenous in structure. The high volume is connected with specific binary data formats and requires more IT-infrastructure, more effort in data utilisation and specific emphasis in bit stream preservation. Another aspect is the selection of data entities which are suitable for long-term preservation. Because of the large amounts of data from numerical models and satellites it is not feasible to move all data into long-term data archiving.

Developments in IT-infrastructure and networks provide the basics to access larger amounts of data from geographically distributed data archives. Based on the Grid-infrastructure first implementations of data federations in Earth system research with special emphasis on model data have been performed. In the UK the National Environment Research Council (NERC) data grid,[3] in Germany the Collaborative Climate Community Data and Processing Grid (C3Grid)[4] and in the USA the Earth System Grid (ESG)[5] have been developed and implemented; the ESG had been chosen as infrastructure to implement the CMIP5/AR5 data federation. The C3Grid architecture has been chosen here as an example. Central to data is their three layer structure and the associated virtualisation of data and compute nodes. At the user interface only services are offered but no direct access to physical resource locations (Fig. 2.2).

The most recent European effort to implement infrastructure to support Earth system modeling is the EU-FP7 project Infrastructure for the European Network for Earth System Modelling (IS-ENES).[6] The data portal is included in an overall dissemination web-interface for Earth system modeling information. The IS-ENES data portal establishes the link to the CMIP5/AR5 data federation and additional data

---

[3] http://ndg.nerc.ac.uk

[4] http://www.c3grid.de

[5] http://www.earthsystemgrid.org

[6] http://is-enes.enes.org

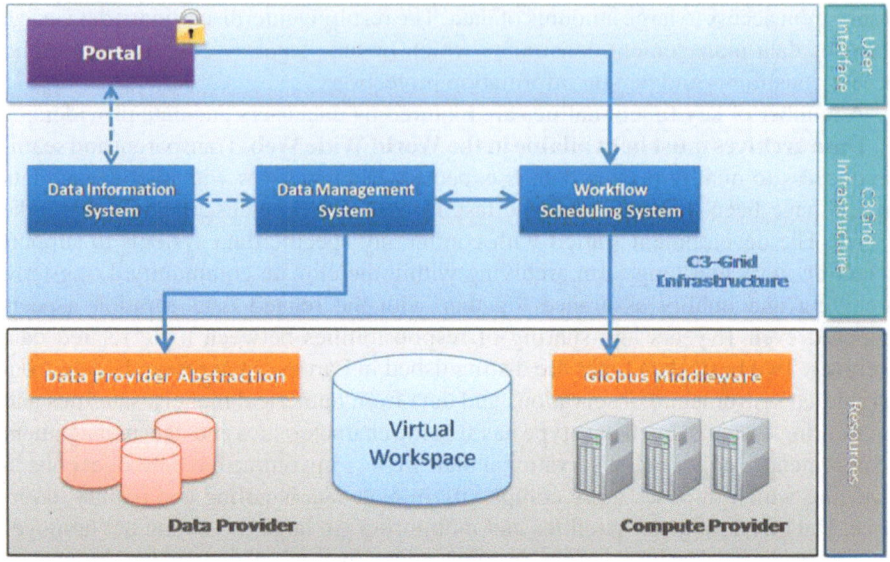

Fig. 2.2 General architecture of C3Grid infrastructure

archives. The data network includes additional compute nodes for server side data processing as in the C3Grid (Fig. 2.3).

Searchable data catalogues must be available which provide information about the archive content. The corresponding metadata are expected to be complete with respect to inter-disciplinary data utilisation.

Interdisciplinary data utilisation requires more complex metadata models than for discipline specific data systems. General discipline specific knowledge cannot be assumed as widely available in the heads of users apart from the data system. This additional information is mostly not included in discipline specific legacy systems.

Data federations which allow for transparent data search and data access across geographically distributed data archives require a common metadata model and semantic mapping in addition to common data access interfaces. The federation metadata model forms the federation data repository (see Data Information System in the C3Grid architecture or the data repository in IS-ENES). The federation metadata model for geo-referenced data in climate research should encourage the interdisciplinary data utilisation which means completeness with respect to the Institute of Electrical and Electronics Engineers (IEEE) metadata reference model (Bretherton 1994). Metatdata for climate research data should support four different interfaces to scientific data management:

- Browse, Search and Retrieve.
- Ingestion, Quality Assurance and Reprocessing.
- Application to Application Transfer.
- Storage and Archive.

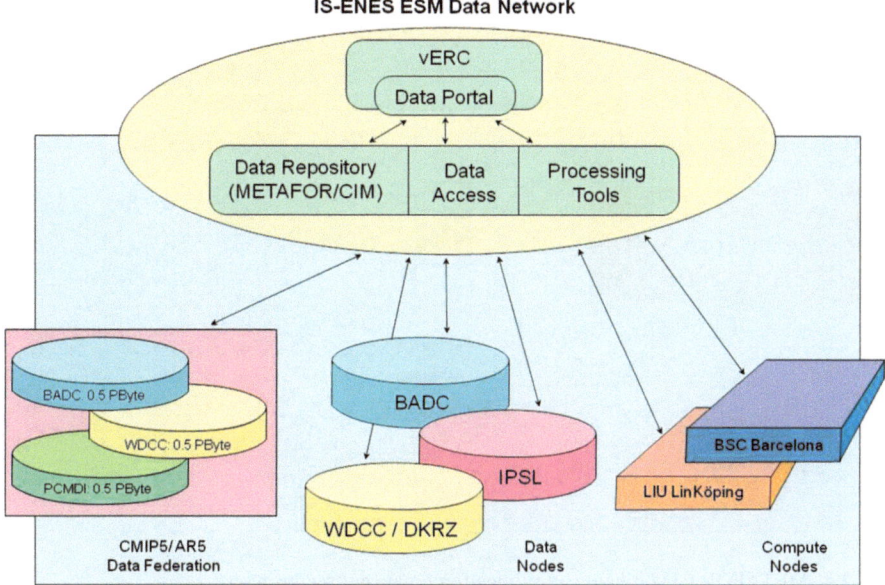

Fig. 2.3 General architecture of the IS-ENES data network which includes data nodes and compute nodes. The data portal is integrated in an overall ESM information portal

General agreement in European federations for geographical data is to make use of ISO 19115 [7] as metadata basis. The EU-FP7 project Common Metadata for Climate Modelling Digital Repositories (METAFOR)[8] has developed a Common Information Model (CIM) with emphasis on Earth system model data. CIM will complete the existing metadata information models by searchable numerical model information and data provenance description. Other aspects of inter-disciplinary data utilisation and data federations are ontologies and controlled vocabularies. Community specific vocabularies have to be matched and must be managed in order to make use of them in repositories of different communities. Especially the creation and maintenance of controlled vocabularies are central development aspects in METAFOR (Fig. 2.4).

Legacy data archives which participate in data federations have to ensure the mapping from their legacy metadata systems to the federation metadata model. Presently this mapping process will be performed by Extensible Markup Language (XML) export and import together with style sheet transformation between the source and the target XML. The target XML can be offered for harvesting in the data federation e.g. by the Open Archives Initiative Protocol for Metadata Harvesting (OAI-PMH).[9] These harvested metadata entities are integrated in a common repository and form the federation data catalogue. The difficulty is the content matching and the trans-

---

[7] http://www.iso.org/iso/catalogue_detail.htm?csnumber=26020

[8] http://www.metaforclimate.eu

[9] http://www.openarchives.org/pmh

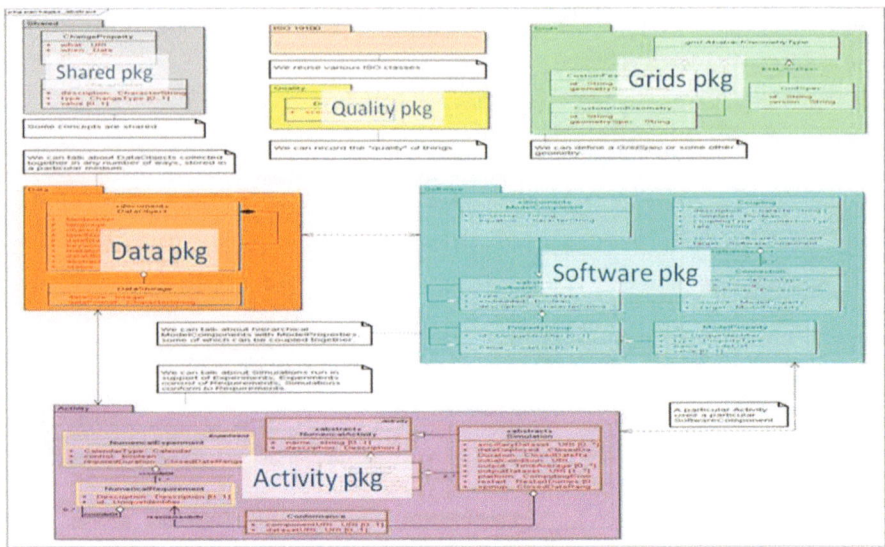

Fig. 2.4 METAFOR CIM schema http://metaforclimate.eu/trac/browser/CIM

formation definition of individual data entities from the legacy data model into the federation data model (ontology).

Server side data processing must be available for data reduction and format conversion. Interdisciplinary data access expects a transition from data access (pure numbers) to information provision. Data archive users request along with the metadata and pure data access also options for data reduction and data manipulation. Data processing requirements depend on the specific user community. Roughly the modelling community and the interdisciplinary data usage can be distinguished. The modelling community would like to see complex workflows which access large amounts of data for analysis of climate model results. Workflow examples are storm track analysis, hurricane detection and multi-model, multi-ensemble averages. Interdisciplinary data usage which is presently related to climate change assessment, mitigation and adaptation requires data reduction and data conversion for server side data processing. Format conversion includes conversion into input for the Geographical Information System (GIS) or ASCII for spread sheet usage. Data reduction includes extraction of geographical sub regions from global fields and interceptions of time series. Sever side data processing additionally includes grid transformation, visualisation and calculation of characteristic, application dependent parameters.

Server side data processing is at the very beginning in federated data systems but will gain more importance since the amount of data grows and interdisciplinary data usage increases. Presently we see searchable federation catalogues and transparent data access in geographically distributed archives. Future development requires a transition from data access to information provision.

Legal aspects and access constraints must be covered by the corresponding security infrastructure. Realisation of the Open Data paradigm with anonymous data access is at the very beginning but it does not seems feasible in all cases. Not all data are available under "Open Access" but most of the scientific data are freely available for academic or non-commercial use. Freely available means free of charge for standard Internet access, additional services like copies to CD/DVD normally are charged at cost basis. Freely available does not mean anonymous data access. Authentication is mostly requested mainly for two reasons: user notification in case of data errors and analysis of archive system usage for documentation and improvement. Additional authorisation beyond the separation of academic and non-academic use is not necessary in most cases. It is only requested for a few data entities from e.g. European weather services.

Another aspect is the "single sign on" requirement in international data federations. The implementation of dedicated security infrastructure for authentication and authorisation is presently not fully solved in most cases. Problems appear due to legal and software aspects. The rules for protection of personal data and copy-rights differ at the national level, even within Europe. Security infrastructures in international data federations have to take national regulations into account. Presently there is no widely accepted standard security infrastructure for data federations available, not even in Europe. Security software is still under development. Approaches under consideration are Shibboleth[10] and OpenID. [11] OpenID is used in commercial applications as well.

Instruments to structure available information must be developed and implemented. Topic related data and information portals are presently developed to provide guidance for specific user communities. Recent developments in Internet technology (Web 2.0) provide infrastructures for web-based, online data access (Fig. 2.5).

Web-based transparent data access requires information on available data and information (metadata). Topic related portals try to present available information with regards to content or with respect to specific applications. Portals are implemented as web-based graphical user interfaces which allow for search and/or selection of information or data entities. Over the years portals showed specific steps in development:

- Portal development started as simple link lists with explanations of the content behind the referenced Uniform Resource Locator (URL).
- The next step realized searchable, domain specific catalogues (metadata) including external links to data archives or specific data entities.
- The third stage in development adds the transparent data access to one or more data archives (federation).
- The next step will open existing domain specific data federations for interdisciplinary data usage. This mainly sets strong standardisation requirements to metadata content, catalogue implementation and search options while data access itself has to be completed by server side data processing.

---

[10] http://shibboleth.internet2.edu

[11] http://openid.net

Elements of the Web's Next Generation

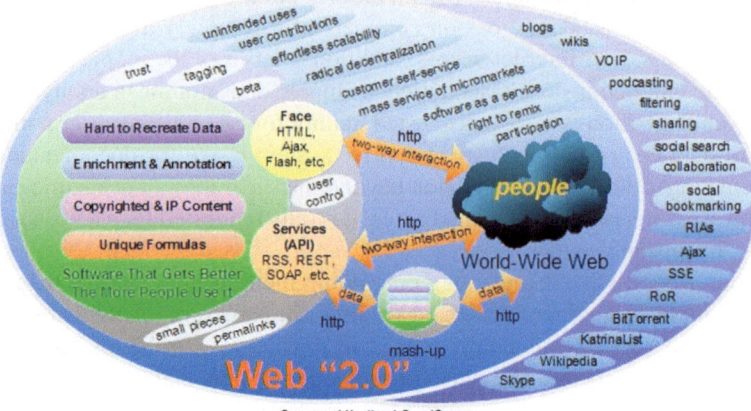

Fig. 2.5 Elements of the web's next generation

Presently the development in ESM is in stage three with transition to step four. Today one of the major drawbacks is the implementation of security infrastructures for user authentication and authorization.

The following sections in this chapter highlight some data and information management aspects in more detail which are already mentioned in the presented overview of ESM data infrastructures.

References

Bretherton F (1994) Reference Model for Metadate: A Strawman.Technical report, University of Wisconsin. URL http://citeseerx.ist.psu.edu/viewdoc/download?doi=10.1.1.48.5381&rep=rep1&type=pdf

Lautenschlager M, Reinke M (1997) Climate and environmental database system. Kluwer Academic Publishers, Boston.

Chapter 3
Harvesting of Metadata with Open Access Tools

Uwe Schindler, Benny Bräuer and Michael Diepenbroek

Search engines like Google have changed the way scientists are searching for publications and data. Users are accustomed to a single input line to enter search terms. They expect fast search response times, ranked results, or recommendation systems as e.g. faceted searches. This not only requires a change from relational databases, which are commonly used in geosciences for data retrieval, to full-text search (FTS) engines (Bennett 2004), but also usage of centralized metadata catalogues.

To create such catalogues, two steps are necessary: Harvesting metadata from data providers and building an index to enable FTS. As example for the implementation of such a tool providing these functionalities the C3Grid project has been chosen. C3Grid[1] uses a generic portal software (panFMP—PANGAEA® Framework for Metadata Portals) for its Data Information Service (DIS). The data providers in the C3Grid generate ISO-19139 conformant metadata for the objects in their databases and file systems, make them available for harvesting by various Open Archives Initiative Protocol for Metadata Harvesting (OAI-PMH) repository software (e.g. Digital Libraray for Earth System Education (DLESE) Java-based open source Open Archives Initiative (jOAI) software).

---

[1] The Collaborative Climate Community Data and Processing Grid (C3Grid) proposes to link distributed data archives in several German institutions for Earth system sciences and to build up an infrastructure for scientists which provides tools for effective data discovery, data transfer and processing.

U. Schindler(✉) · M. Diepenbroek
MARUM—Zentrum für Marine Umweltwissenschaften, Bremen, Germany
e-mail: uschindler@pangaea.de

M. Diepenbroek
e-mail: mdiepenbroek@pangaea.de

B. Bräuer
Alfred-Wegener-Institut für Polar- und Meeresforschung, Bremerhaven, Germany
e-mail: Benny.Braeuer@awi.de

W. Hiller et al., *Earth System Modelling – Volume 6*, SpringerBriefs in Earth System Sciences, DOI: 10.1007/978-3-642-37244-5_3, © The Author(s) 2013

3.1 Background and Motivation

During the last decade there had been various initiatives and approaches in networking global data services. A main focus had been on Earth sciences and related data. The lately implemented *Group on Earth Observations* (GEO) conceived a plan for a sustained *Global Earth Observations System of Systems* (GEOSS, Battrick 2005) which largely builds on concepts of Global Spatial Data Infrastructures (GSDI).

Correspondingly, the EU directive *Infrastructure for Spatial Information in Europe* (INSPIRE, The European Parliament and Council 2007) ensures that the spatial data infrastructures of the Member States are compatible and usable in a community and transboundary context.

These two initiatives clearly emphasize the need for portal frameworks to provide simple and transparent access to geoscientific data and metadata. In fact, global standardization efforts have been quite successful in recent years, leading to some convergence in developments. Nevertheless, portal developers have to cope with two problems: First, the fact that data providers have different backgrounds and capabilities and furnish data and metadata using various protocols and content structures. In practice, puristic, homogeneous approaches based on unique standards are likely to cause exclusion of at least part of the potential data sources and—due to the dynamics in SDI development—might result in a premature end of operation. Grid approaches, however, in general follow heterogeneous approaches thus avoiding this problem. Second, the above-mentioned Google-like search expectations.

Sophisticated search functionalities on metadata are not part of current Grid concepts. Common portal frameworks do not support metadata retrievals. Distributed catalogues in a Grid environment are mostly used for control of workflows but not for data discovery.

3.2 The Metadata Portal Framework "panFMP"

C3Grid uses panFMP[2] (Schindler and Diepenbroek 2008), a generic and flexible framework based on Apache Lucene (Hatcher et al. 2009) for building geoscientific metadata portals independent of content standards for metadata and protocols (Fig. 3.1). Metadata from data providers can be harvested with commonly used protocols (e.g. OAI-PMH) and metadata standards like ISO-19139. For this purpose the Java-based portal software supports any Extensible Markup Language (XML) encoding and makes metadata searchable through Apache Lucene. Software administrators are free to define searchable fields independent of their type using XPath. In addition, by extending the full-text search engine Apache Lucene, panFMP has significantly improved the user interaction with the portal. It is now possible to issue queries for numerical and date/time ranges by supplying a new trie-based algorithm, thus enabling high-performance space/time retrievals in

---

[2] PANGAEA® Framework for Metadata Portals. http://www.panfmp.org

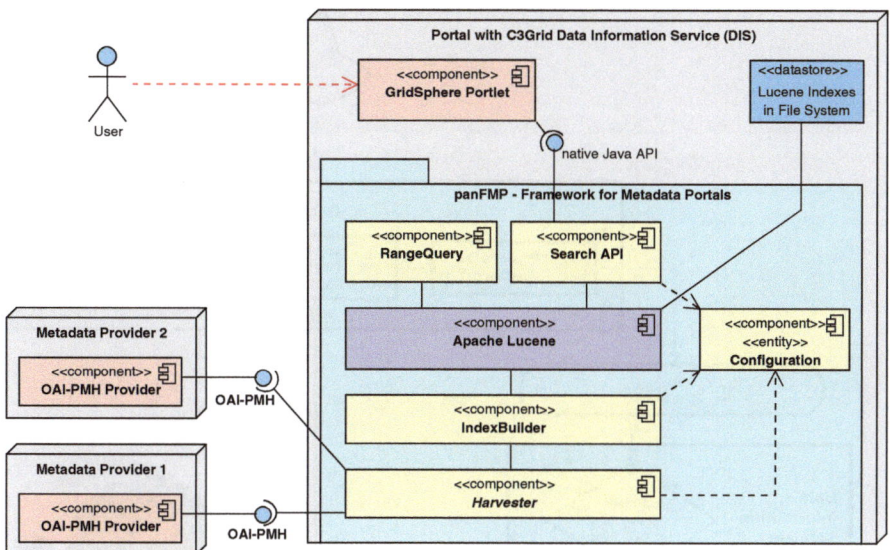

Fig. 3.1 Structure of panFMP

full-text search based geo portals. As data resources in C3Grid are related to geolocations this feature has proven to be very helpful. The trie-based code was donated to the Apache Software Foundation and included into version 2.9 of the software.

The portal-specific Java API and web service interface is highly flexible and supports custom front-ends for users, provides automatic query completion (Asynchronous JavaScript and XML—AJAX), and dynamic visualization with conventional mapping tools. The panFMP software was made freely available through the open source concept.

3.3 Adoption of the Framework in C3Grid

The panFMP metadata framework as core of the Data Information Service (DIS) is a central component of C3Grid. From the indexes created with panFMP, all necessary information is provided that is needed to create job descriptions for data download requests and workflow submission. Therefore, the framework has an important role for the functionality of the grid.

The DIS consists of two interfaces (Fig. 3.2): (1) the harvesting interface to the data providers—and (2) the query interface.

Harvesting data providers. The data providers supply metadata in ISO-19139 compatible XML format through the OAI-PMH protocol (Van de Sompel et al. 2004). Most data providers have installed the DLESE jOAI software, because it can provide

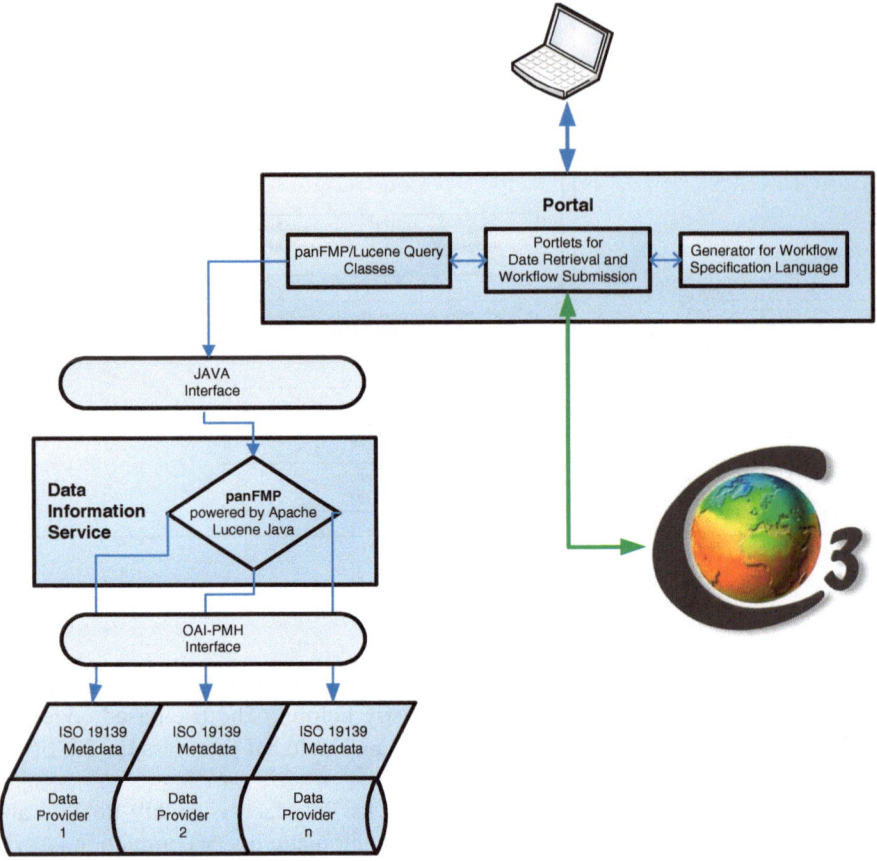

Fig. 3.2 Embedding panFMP into the C3Grid data information service and the GridSphere portal

various metadata formats from already prepared XML files. For C3Grid, the metadata
format was fixed to ISO-19139 (XML encoding of ISO-19115, see also, Kresse and
Fadaie 2004), but because of the flexibility of panFMP, other formats like the Data
Interchange Format (DIF)[3] of Global Change Master Directory (GCMD) would
also be possible. This means that because of the on-the-fly metadata transformation
abilities of panFMP, the integration of metadata from other grids is possible without
too many problems.

panFMP can handle large volumes of metadata. Nevertheless, in C3Grid the size
of metadata records and corresponding fulltext indexes is small (megabytes) compared to the related data volumes (terabytes). Initiated through a cron-job running
twice a day, metadata from data providers is harvested incrementally. In parallel the
corresponding indeces are updated.

---

[3] http://gcmd.nasa.gov/User/difguide

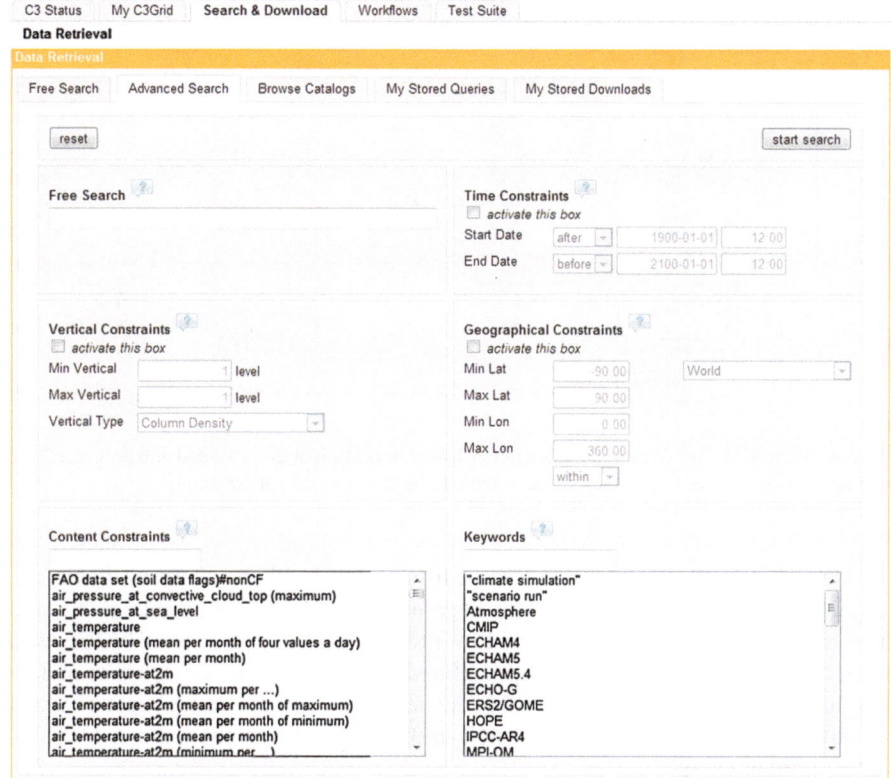

Fig. 3.3 Full-text search interface of the C3Grid portal

Query interface. The second interface of panFMP can be used to query the metadata indexes. panFMP provides a Simple Object Access Protocol (SOAP) based interface to the catalogue and also a Java API. As the search engine is directly running inside the GridSphere Portal Framework, the way through SOAP, which is more limited, was not chosen. Instead, the DIS portlets directly access panFMP using the Java API.

The user has the ability to search all metadata fields simultaneously as well as searching specific fields (Fig. 3.3). For the latter various forms are provided, which allows users to select parameters, enter bounding boxes, date/time ranges, and vertical coverages. The metadata retruned are presented structurally and offer users various possibilities to work with related data. The different supported workflows are shown in the Unified Modeling Language (UML) activity diagram in Fig. 3.4:

Users can subselect single data sets as a whole or define data fractions to be cut out and create corresponding requests to the data provider's data access interface. These requests are embedded into a C3Grid Workflow Specification Language (WSL) documents, which are based on Job Submission Description Language (JSDL, see Anjomshoaa et al. 2005). Subsequent processing of requests is handled by the

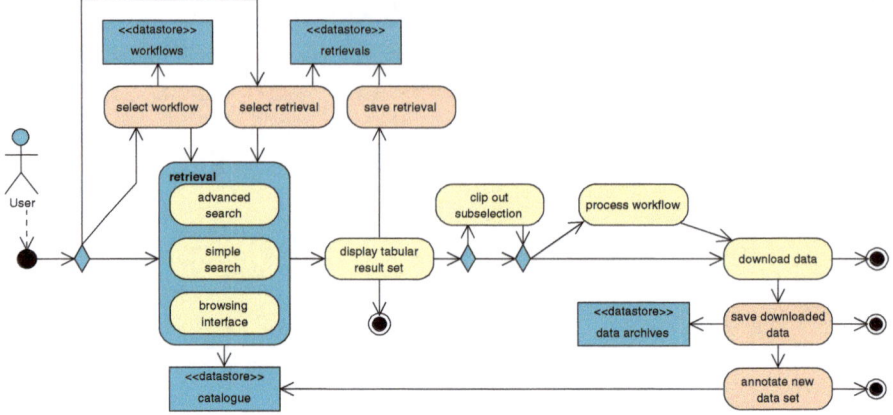

Fig. 3.4 Search capabilities on metadata in C3Grid. Annotation of downloaded and processed data with new metadata was not yet fully implemented *(right lower side of diagram)*

Workflow Scheduler Service of C3Grid. It is also possible to query available data sets, that fit a specific workflow. This becomes possible through predefined search queries, which were added to the workflow properties and are processed by the workflow portlets (Fig. 3.5). The lists of predefined data sets can be very long, so that the user may loose the overview. For that, in the new version of the portal, facetted search is available and the possibility to search in the lists to lower the number of results will be implemented.

Whenever a user issues a specific workflow, the workflow scheduler extracts the data retrieval constraints from the WSL document and calls the data provider's data access interface, which extracts the data from the local data inventory (e.g. by doing Structured Query Language (SQL) to the underlying database or data warehouse, or extracting subsets of data from larger file repositories). Subsequently, retrieved data sets are transferred to a free compute node which—according to the descriptions in the WSL file—takes over the final processing of the data. If the user just wants to download the extracted data, the file can be transferred directly to the data portal and downloaded to the user's browser.

A publishing workflow is under development to re-integrate results into the grid again. For that, the data providers are defining new OAI-Sets or, if necessary, setup new OAI-servers to separate these metadata from the original ones. To avoid a growing amount of data which is maybe only used once by one user, the data and the metadata is purged after 24 h.[4]

---

[4] a lengthening is possible.

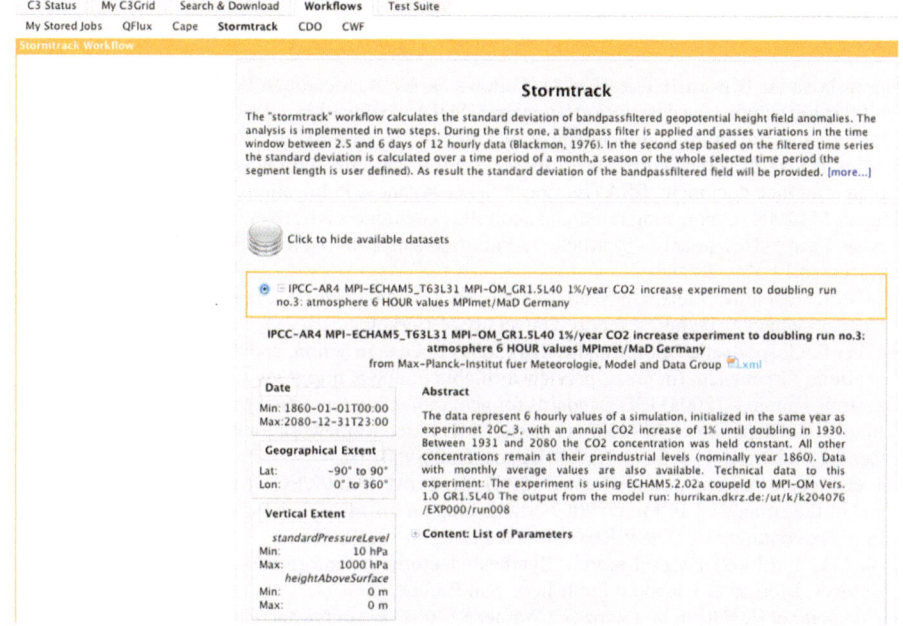

Fig. 3.5 Selecting available datasets for the "stormtrack" workflow. The query is run in panFMP by asking for datasets matching the requirements of the workflow

3.4 Conclusion

panFMP could seamlessly be integrated into C3Grid. Search performance and capabilities supplied through the Apache Lucene based panFMP allows for efficient and scalable metadata retrievals. The generic implementation of panFMP enables extension to other application fields within grid networks and thus further propagation of grid architectures—or their successors, the cloud.

With this approach it is much easier for the scientific user to find and retrieve distributed data within the C3Grid portal. The user benefits from a performant search and a mighty query syntax which can be used to build sophisticated search forms (e.g. Fig. 3.5). The grid portal allows the user to search for datasets by full text, variable names, date/time constraints or a bounding box and start jobs on selected datasets. In the future, the C3Grid search engine and the underlying panFMP will be extended to also support faceted navigation, allowing the user to further drill down the search results based on criterias selected from a classification system (English et al. 2002; Tunkelang 2009). panFMP will take advantage of the recently added faceting module of Apache Lucene.

Usage of the panFMP software is in contrast to the conventional grid approach where the search engine would be distributed to all data providers, but has important advantages, described in this chapter, e.g. the removal of difficulties of FTS engines with numerical and date searches (Schindler and Diepenbroek 2008).

References

Anjomshoaa A, Brisard F, Drescher M, Fellows D, Ly A, McGough S, Pulsipher D, Savva AE (2005) Job submission description language (JSDL) specification, version 1.0. Global grid forum. http://www.gridforum.org/documents/GFD.56.pdf. Accessed 17 Jul 2009

Battrick B (2005) Global earth observation system of systems (GEOSS) 10-year implementation plan reference document. ESA (European Space Agency) Publications Division, Noordwijk

Bennett M (2004) Contrasting relational and full-text engines. NIE (New Idea Engineering) Enterprise Search Newsletter 2(9):article 1. http://ideaeng.com/pub/entsrch/issue09/article01.html. Accessed 14 May 2009

English J, Hearst M, Sinha R, Swearingen K, Yee KP (2002) Flexible search and navigation using faceted metadata. Technical report. University of Berkeley

Hatcher E, Gospodnetić O, McCandless M (2009) Lucene in action, 2nd edn, p 475. Manning Publications, Greenwich. (in press, preview available online at http://www.manning.com/hatcher3/)

Kresse W, Fadaie K (2004) ISO standards for geographic information. Springer, Heidelberg 322 pp

Schindler U, Diepenbroek M (2008) Generic XML-based framework for metadata portals. Comput Geosci 34(12):1947–1955. doi:10.1016/j.cageo.2008.02.023

The European Parliament and Council (2007) Directive 2007/2/EC of the European parliament and of the council of 14 March 2007 establishing an infrastructure for spatial information in the European community (INSPIRE). Official J Eur Union 50:1–14

Tunkelang D (2009) Faceted search. Synthesis lectures on information concepts, retrieval, and services, Morgan & Claypool Publishers, San Rafael

Van de Sompel H, Nelson M, Lagoze C, Warner S (2004) Resource harvesting within the OAI-PMH framework. D-Lib Magazine 10(12)10.1045/december2004-vandesompel

Chapter 4
Data Discovery: Identifying, Searching and Finding Data

Stephan Kindermann

Data analysis, data inter-comparison and data assimilation are at the heart of any Earth system science workflow. Aditionally, Earth system data are used in many interdisciplinary research activities and climate impact studies. Thus the ability to effectively locate and obtain datasets based on specific data characteristics and properties is crucial for scientific progress. Data discovery mechanisms are required to support users in locating data in internationally distributed data centers and get contact to the appropriate data access and data processing services. Data which is not discoverable is lost from a users perspective.

   This chapter provides an overview of data discovery in the context of ESM data archives. After the definition of some vocabulary (Sect. 4.1) and an overview of ESM data discovery challenges (Sect. 4.2) a generic architectural model is defined (Sect. 4.3). To build up a discovery architecture for world wide distributed ESM data archives the adoption of wide spread standards is a prerequisite. Thus important standardization efforts of the individual components of the generic architectural model are summarized in Sect. 4.4. Finally, state of the art efforts to establish uniform data discovery architectures for ESM data are characterized in Sect. 4.5.

4.1 Overview and Definition of Terminology

On the one hand, Earth system data repositories hold multidimensional, high volume ("gridded") data sets, which contain information defined on structured n-dimensional arrays (grids) normally resulting from model runs. For example, data volumes in ongoing climate model intercomparison projects like CMIP5[1] are in the multi Petabyte range.

---

[1] http://cmip-pcmdi.llnl.gov/cmip5

S. Kindermann (✉)
Deutsches Klimarechenzentrum, Hamburg, Germany
e-mail: kindermann@dkrz.de

W. Hiller et al., *Earth System Modelling – Volume 6*, SpringerBriefs in Earth System Sciences, DOI: 10.1007/978-3-642-37244-5_4, © The Author(s) 2013

On the other hand, typical scientific workflows also include access to various types of observation and measurement data e.g. for model result validation or model parametrization. Aditionally, scientific data analysis activities produce new, derived data sets. All these data products coming from various sources have to be described consistently by detailed metadata to be discoverable and to give users confidence in data origin and integrity. Besides the challenges imposed by the amount of data and the heterogeneity of its content as well as complex data provenance, data normally is maintained in separate administrative and policy domains and is only described by locally defined annotations.

Thus the first step enabling uniform data discovery across separate data centers, is the definition of standardized descriptive information (*metadata*) about data (and data services). This information is published to *catalogues* exposing discovery services which *discovery portals* or other applications can contact based on standardized *query protocols* and languages. This generic model is depicted in Fig. 4.1 and described in more detail in Sect. 4.4.

Following INSPIRE drafting team "Network Services" (2008a) we will use these definitions throughout this chapter:

- *Discovery:* The inquiry of the nature and content of a spatial resource.
- *Discovery Service:* Distinct part of the functionality that is provided by an entity through interfaces for the inquiry of the nature and content of a spatial resource.
- *Metadata:* Information describing spatial datasets and spatial data services and making it possible to discover, inventory and use them.
- *Spatial Resource:* Asset or means that fulfills a requirement and has a direct or indirect reference to a specific location or geographic area; example: dataset, service.

4.2 ESM Data Discovery Challenges

In this chapter a set of high level ESM data discovery challenges are summarized. This characterization is later used in Sect. 4.5 to evaluate how recent ESM data discovery infrastructures respond to these challenges.

4.2.1 Data Product Granularity and Aggregation

ESM data products are mainly available in files based on self describing data formats (like GRIB, NetCDF, HDF5) containing basic "use level" metadata, sometimes constrained by vocabularies. Normally data products are stored and described in aggregations. Thus the definition of aggregation layers strongly influences the basic functionality which can be offered to users by data discovery services. Typical dimensions defining aggregation layers are: time series, variables, experiments, models. Thus e.g. the aggregation convention used in CMIP5 is defined in a model data output requirement document[2] and defines *chunks* of files describing single physical *variables* grouped along *time* produced by *experiments* using certain *models*.

4.2.2 Data Provenance and Scientific Context

To evaluate the appropriateness of data for their specific analysis tasks, scientists need to discover properties of data provenance as well as the scientific context in which the data was produced. Making available this information is complex since usually dedicated *data curators* are required to collect and ingest this information into metadata repositories and link it to data products. Tools can sometimes help to automatically add provenance information like e.g. concrete information on the models which generated the data along with their configuration and parametrization characteristics. A generic provenance and scientific context description should be based on a sound (e.g. ontological) model to characterize the semantic context of a concrete data product.

4.2.3 Data Lifetime and Quality

ESM data produced by model runs are subject to change. Errornous data has to be removed, new versions of existing data products are generated as a result of e.g. model run or post processing chain modifications. Especially new data having a high probability of future changes is of high interest to users. On the other hand data quality can only be assured stepwise by following a well defined sequence of quality assurance steps. Data quality information as well as persistent identifiers assigned to data products are of high interest to users wishing to publish research results e.g. in (electronic) journals.

4.2.4 Data Volume and Data Distribution

Nowadays ESM data intercomparison projects work on Petabytes of model output data. Automatic tools and well defined workflows are necessary to generate and publish consistent metadata descriptions. Additionally, the data is not centrally stored, but available at geographical and organizational separated data centers. Users want to discover related data products at an abstract, consistent logical level which hides the distributedness and heterogeneity of the data landscape. This requires

---

[2] CMIP5 Model Output Requirements: File Contents and Format, Data Structure and Metadata: http://cmip-pcmdi.llnl.gov/cmip5/docs/CMIP5_output_metadata_requirements.pdf

the establishment of metadata agreements and the deployment of a consistent and scalable data discovery middleware layer for ESM data providers.

4.2.5 Interdisciplinarity and Heterogeneity of Data Discovery Use Cases

ESM data is used heavily for climate impact studies and constitutes one data source among others in many interdisciplinary research activities. Data discovery expectations from users are very heterogeneous. Some expect detailed and sophisticated discovery possibilities whereas others expect simple and pre-defined discovery routes. In general dedicated portals for specific usage scenarios are necessary. These portals should (re-)use common, standardized basic search and discovery services provided by the metadata middleware, requiring an overall modular data discovery architecture.

4.2.6 Search Scalability and Flexibility

Different technological and infrastructural approaches exist to build up search services scaling to hundreds of millions of ESM data products which are grouped in various aggregations. One main characteristic is the distribution of search functionality ranging from centralized approaches (e.g. large relational databases or index stores) to fully distributed approaches. Keeping centralized stores up to date based on metadata harvesting is difficult, whereas in a distributed scenario with search results aggregated from many decentralized instances stable search results are difficult to guarantee.

4.3 Generic Data Discovery Architecture

An overview of a generic ESM data discovery architecture is given in Fig. 4.1, which is inspired by the architectural model outlined by INSPIRE drafting team "Network Services" (2008b). Spatial data is stored in data archives based on different technologies (e.g. databases, data-warehouses, file systems, tape robots). These archives in general expose services to view datasets, download datasets and pre-process datasets:

- *view services* make data accessible to e.g. map viewer clients. They in general provide information like visual representations of data content based on well known coordinate reference systems (CRS).
- *download services* make datasets available based on different data access protocols (e.g. ftp, http, ..) using different authentication and authorization approaches.

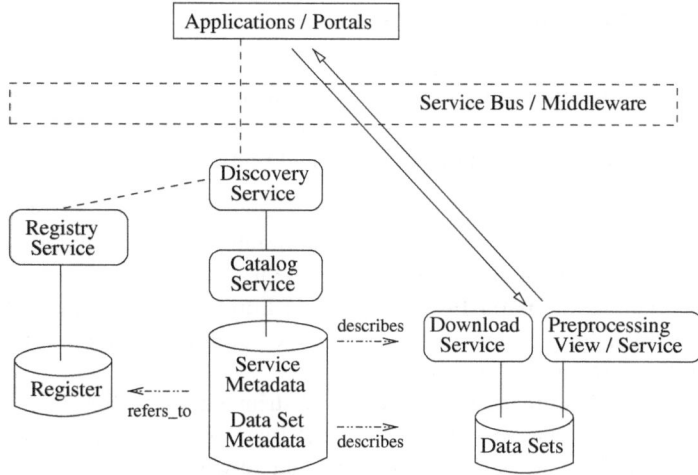

Fig. 4.1 Generic data discovery architecture

- *preprocess services* allow for server side data size reduction and data filtering before data delivery. This is of specific interest for high volume ESM data access.

All data sets as well as data services are annotated by *descriptive metadata* which is available in *catalogues* and freely accessible for *discovery services*.

Additional services at an infrastructure level can provide processing capabilities to operate on data sets, providing building blocks for complex data analysis workflows. In general some of these processing capabilities are co-located with the data centers (e.g. server-side analysis and subsetting) to allow for data reduction before data delivery.

To allow for transparent discovery of heterogeneous data sets from different contexts (e.g. observation data and model output data) additional information resources need to be maintained properly and made available in *registers* (via *registry services*). Based on a well defined governance model registry contents are normally associated to unique, unambiguous and permanent identifiers and try to define a very basic vocabulary to describe entities of interest. Examples of such registers are:

- *feature catalogues* containing definitions and descriptions of spatial object types
- *code lists* describing attribute values and their domains or
- *thesauri* with additional interrelation info (like hierarchies, etc.)
- *references* like coordinate reference systems or units of measurements

Registries are the base ground to enable the buiding of frameworks supporting semantic interoperability.

4.4 Implementation Standards

Important standardization efforts for the components described in the architectural model are either driven top down by organizations and initiatives or bottom up by heavily used "de facto" developments. Important representatives are:

- The Open Geospatial Consortium (OGC),[3] which defines interoperable catalogs based on a common abstract model and common interface model, supporting discovery, catalog maintenance and management.
- The Infrastructure for Spatial Information in the European Community (INSPIRE)[4] initiative, which defines guidelines enabling the development of interoperable services for geo-referenced data products.
- The Thematic Realtime Environmental Distributed Data Services (Thredds) project[5] as well as the open Source Project for a Network Data Access Protocol (OPeNDAP)[6] framework for scientific data networking.

4.4.1 Metadata Standards and Initiative:

Often adopted minimal and generic standards for discovery metadata are the Directory Interchange Format (DIF)[7] and Dublin Core.[8] A much more expressive and detailed metadata standard for Geospatial metadata in provided by ISO 19115[9] with its ISO 19139[10] XML representation. INSPIRE promotes the use of ISO 19139 for data metadata and ISO 19119 for service metadata description. The open geospatial consortium (OGC) adopts the ISO standards by providing ISO profiles for OGC services.

On the other hand a de facto standard which is often deployed in ESM data environments is Thredds. It provides data discovery middleware components enabling scientific data discovery and use. Tools are available to automatically extract use level metadata from e.g. NetCDF file headers and generate Thredds discovery metadata catalogs. Also, recently additional tool support is available for the translation of discovery elements into the ISO 19115 standard (ncISO[11]).

---

[3] http://www.opengeospatial.org

[4] http://inspire.jrc.ec.europa.eu

[5] http://www.unidata.ucar.edu/projects/THREDDS

[6] http://opendap.org

[7] http://gcmd.nasa.gov/User/difguide/whatisadif.html

[8] Dublin Core Metadata Initiative: http://dublincore.org/

[9] ISO 19115:2003: http://www.iso.org/iso/catalogue_detail.htm?csnumber=26020

[10] ISO/TS 19139:2007: http://www.iso.org/iso/catalogue_detail.htm?csnumber=32557

[11] http://www.ngdc.noaa.gov/eds/tds

Standardization efforts for describing ESM data including their general context (e.g. climate model characteristics) were funded in the US Earth system Curator project Curator (2008) and the European Common Metadata for Climate Modelling Digital Repositories (METAFOR)[12] project. METAFOR is creating a Common Information Model (CIM) for climate data and the models that produce it. These efforts attempt to establish a complete data description framework for model output data, including their dependencies on model components, their coupling methods and their configuration. CIM metadata for the CMIP5 project is collected in an online questionnaire[13] and it can be harvested by portals using web feeds.

The availability of discovery services based on these complex metadata descriptions will enable a new generation of data discovery portals, supporting users in finding data based on scientific context information.

4.4.2 Metadata Exchange and Discovery Protocol Standard

A very simple and often widely deployed mechanism to publish discovery metadata is based on the Open Archives Initiative Protocol for Metadata Harvesting (OAI-PMH)[14] protocol. This http-based mechanism is independent of the XML representation of the metadata to be published and many ready-to-use tools and libraries exist (e.g. provided by the Digital Library for Earth System Eduction (DLESE)). A much more complex and general approach is standardized in the OGC Catalogue Service specification for the Web (CSW), supporting discovery, evaluation and use of spatial resources. Adoption to the needs of particular domains is provided through application profiles. Application profiles (AP) which are standardized so far include

- a CSW ISO AP based on a ISO 19139 conform encoding of ISO 19115 and ISO 19119
- a CSW Electronic Business Registry Information Model (ebRIM) profile based on the ebXML Registry Information model of the Organization for the Advancement of Structured Information Standards (OASIS) consortium
- a profile for use by the Earth Observation Community (draft status)

Besides these most prominent metadata exchange standards, also the International Standard Maintenance Agency Z39.50[15] search and retrieval protocol is oftentimes used and the Federal Geographic Data Committee (FGDC)[16] has developed a specific profile for geospatial metadata.

---

[12] http://metaforclimate.eu

[13] http://q.cmip5.ceda.ac.uk

[14] http://www.openarchives.org/pmh

[15] http://www.loc.gov/z3950/agency

[16] http://www.fgdc.gov

4.4.3 Standardization of Registries

The standardization process in the context of ESM data related registries is at its beginning. The Climate and Forecast(CF) metadata convention[17] provides an important basis to enable the processing and sharing of ESM data files created by the NetCDF API.

Registry content in general is structured based on concept schemes, which are domain specific and clearly beyond generic standardization efforts like INSPIRE and OGC. Nevertheless a simple but powerful generic concept description approach is available: the Simple Knowledge Organization System (SKOS).[18] Simple relationships between concepts (e.g. "is-narrower then, is-broader then") can be stated based on Resource Description Framework (RDF)/XML. These relationships can be easily integrated in XML schema definitions, thus allowing to maintain glossaries along with the definition of their entries.

Servers used in production in the context of ESM data include the Natural Environment Research Council (NERC) vocabulary server[19] and the Geonetwork catalog[20] (see also Sect. 4.5).

4.5 Data Discovery Architectures in the Times of the Grid

Based on the previous coarse classification of components and standards, in the following section we will provide a short overview of recent approaches to data discovery in distributed and federated ESM data archives, which try to provide to users a consistent "data grid" experience.

4.5.1 NERC Data Grid

The Natural Environment Research Council (NERC) data grid project has built an environment to ease data discovery and data access to a large variety of data archives. A data portal is based on a discovery service, which harvests metadata descriptions from catalogs maintained by the archives. The discovery metadata is based on a proprietary schema representing metadata objects for linking environmental sciences (MOLES). This schema can be exported in commonly used formats like DIF, Dublin Core or ISO 19139. The harvesting of the exposed metadata information is based on the OAI-PMH protocol.

---

[17] http://cf-pcmdi.llnl.gov/documents/cf-conventions/latest-cf-conventions-document-1

[18] http://www.w3.org/2004/02/skos

[19] http://www.bodc.ac.uk/products/web_services/vocab

[20] http://geonetwork-opensource.org

An NDG vocabulary service manages lists of standardized terms (thesauri) and provides flexible means for lookup, to e.g. support the uncovering of the meaning of individual entities. The terms are related in hierarchies based on SKOS descriptions. Data access services in NDG are based on more concrete metadata model, the Climate Science Modeling Language (CSML)[21] information model. CSML is Geography Markup Language (GML)[22] based and defines datasets according to frameworks provided by the OGC.

4.5.2 C3Grid

Discovery in the Collaborative Climate Community Computing and Data Grid (C3Grid) project is mainly based on ISO 19115/19139 metadata. Data providers provide OAI-PMH servers for their ISO data descriptions, which are harvested into a central catalog. The generation of ISO metadata is in the responsibility of the individual data provider, nevertheless automatic metadata generation tools were developed to support this complex process. A Lucene based discovery service is exposed and used by the C3Grid portal. Registers provide controlled vocabularies for data content, based on the climate and forecast (CF) convention as well as code lists and coordinate reference system (CRS) information. Additionally, the C3Grid Portal interacts with search services of the Earth System Grid Federation (ESGF) to provide users a possibility to search for data stored in the ESGF based CMIP5 data federation.

Metadata on available data access services is collected in a C3Grid information service based on harvested service metadata. High level data analysis services are available through the C3Grid portal, which are realized based on grid jobs executed in the distributed C3Grid environment. The medatata for these services is given based on XML workflow descriptions of the necessary data staging and processing steps. This workflow is executed by the C3Grid scheduling system in cooperation with the data management system operating on a distributed data cache, where intermediate data products are managed.

The ISO standard based discovery architecture of C3Grid enables data providers to be visible in various portals. Thus C3Grid data providers can e.g. be directly harvested from Geonetwork portals. Geonetwork[23] provides a standards-based open source catalog application, providing a building block to build standards-based ESM data discovery portals. It was developed by the United Nations Food and Agriculture Organization (UN FAO). Multiple metadata exchange protocols (OGC-CSW, OAI-PMH, Z39.50) are supported and metadata editing facilities are integrated (e.g. supporting ISO 19139) in addition to SKOS based vocabulary definition and (distributed) search facilities.

---

[21] http://csml.badc.rl.ac.uk

[22] OpenGIS GML Encoding Standard, http://www.opengeospatial.org/standards/gml

[23] http://geonetwork-opensource.org

4.5.3 Earth System Grid (ESG)

Within the Earth System Grid system discovery metadata used in ESG portals is mainly based on information exposed by Thredds catalogs of the distributed ESGF data nodes.[24] To enable uniform discovery of data entities described in the Thredds catalogs, a standard naming scheme (the so called *Date Reference Syntax (DRS)*) was defined and a standard required (NetCDF/-CF based) data layout was established.[25] Automated tools were developed to support these conventions, especially CMOR2.[26] The automatic generation of Thredds catalogs for CMOR2 compliant NetCDF files is enabled by a tool called the ESG data publisher,[27] which is part of the ESGF data node software.

ESGF Thredds metadata information is combined with higher level CIM metadata characterizing the (coupled) models and experiments which produced the data. This information is collected from atom feeds exposed by the METAFOR questionnaire. In the future a dedicated METAFOR search and discovery service will be available and included into the ESGF infrastructure.

Two variants of ESG search and discovery facilities were developed. The first is based on a hierarchical model based on data nodes which publish metadata to a set of (synchronized) portals. The newer ESGF approach builds on a peer to peer (P2P) model, allowing specific ESGF data nodes to also provide and expose discovery and search facilities for themselves and a set of related peers. The first generation of the CMIP5 distributed archive was built using the classical model where portals translated metadata from data nodes to an internal (RDF based) representation, which was used as backend for the faceted search interface. Scalability issues made it necessary to redesign the faceted search to also use internal (e.g. Lucene[28] based) metadata index stores. The ESGF peer to peer search is based on a combined distributed and local metadata search approach using the Solr[29] technology: Local and Harvested Thredds catalogues are indexed and exposed via RESTful[30] search services.

To be able to assign persistent Digital Object Identifier (DOI)[31] to CMIP5 datasets, which can be cited in e.g. scientific publications, a consistent three stage quality assurance process was defined[32] to check data and metadata quality and consistency.

---

[24] http://esgf.org

[25] CMIP5 Model Output Format and Metadata Requirements: http://cmip-pcmdi.llnl.gov/cmip5/output_req.html#metadata

[26] Climate Model Output Rewriter (CMOR): http://cmip-pcmdi.llnl.gov/cmip5/output_req.html#cmor

[27] ESG publication scripts, http://www2-pcmdi.llnl.gov/Members/bdrach/.personal/esg-publication-scripts

[28] Apache Lucene search engine: http://lucene.apache.org/core

[29] Apache Solr search platform: http://lucene.apache.org/solr

[30] Representational State Transfer (REST)

[31] http://www.doi.org

[32] https://redmine.dkrz.de/collaboration/projects/cmip5-qc/wiki

4.5.4 The European Distributed ESM Data Archive

As part of the IS-ENES FP7 project the European part of the distributed ESGF archive is established, focusing on IPCC AR5 long term data archival as well as the extension to new data products, like regional ESM data produced within the COordinated Regional climate Downscaling Experiment (CORDEX)[33] experiment. Besides the data discovery mechanisms provided by the ESGF infrastructure, IS-ENES developed an additional centralized catalog and search service (implemented in a relational database), based on Thredds metadata harvested from the ESGF data nodes. The search interface is embedded in the ENES portal[34] which will provide an information and service platform for ESM models and data in Europe.

4.6 Summary

Building user friendly data discovery infrastructures integrating distributed ESM data archives has to start with agreements on vocabularies, metadata formats and exchange protocols. The complexity of data generation workflows increases as well as the amount and complexity of generated ESM data products, thus fully automatic metadata generation steps have to be included into the data production process in the future.

To support uniform discovery across ESM data produced within different experiments a standardized, well defined and interoperable metadata handling architecture and workflow has to be defined. Metadata handling components (search indices, metadata generation facilities, metadata providers and harvesters, GUIs) have to be established, maintained and provided sustainably at distributed data centers around the world. Whereas the needed basic technology and standardizations are available, tackling the complexity of defining, establishing and maintaining a consistent distributed ESM data discovery architecture at a global scale is at its beginning.

A first step is currently being exercised within the global ESGF CMIP5 data federation. Starting with well defined experiment definitions, file metadata and file aggregation conventions are established and supported by automated tools like CMOR2 and the ESG data publisher, generating standardized Thredds metadata catalogs. Additionally, the METAFOR questionnaire is used to collect higher level metadata and an overall data and metadata quality assurance procedure was defined. A complete new software stack was developed and has to be established to support this large ESM data intercomparison project.

On the other hand the C3Grid project has its focus on integrating legacy data centers (also ESGF data nodes in the future) and providing a distributed collaborative workspace to support climate data analysis workflows, thus enabling new possibilities for data sharing and data reuse among scientists. A key component of C3Grid are ISO

---

[33] http://wcrp.ipsl.jussieu.fr/SF_RCD_CORDEX.html

[34] http://verc.enes.org

19139 metadata records not only describing published data maintained at C3Grid data centers, but also describing derived data products with their provenance information. Workflows are formally described by an XML based workflow description, enabling the definition of reusable blocks of functionality.

The enabling of distributed data analysis activities using consistently managed distributed ESM data archives is a key requirement which also drives the establishment of a future consistent ESM data discovery infrastructure. ESGF, C3Grid and IS-ENES (with METAFOR) are key projects contributing to this effort.

References

INSPIRE drafting team "Network Services" (2008a) Draft implementing rules for discovery services (IR3), Document. http://inspire.jrc.ec.europa.eu/reports/ImplementingRules/network/D3.7_IR3_Discovery_Services_v3.0.pdf

INSPIRE drafting team "Network Services" (2008b) INSPIRE Network services architecture—Version 3.0, 2008-07-19. http://is.gd/8949v

Curator (2008) Earth system curator: spanning the gap between models and datasets. Project website. http://www.earthsystemcurator.org

Chapter 5
User Driven Data Access Mechanisms

Martina Stockhause

Most climate and climate impact studies include data analyses and therefore rely on scientific data. Generally, only a small part out of the whole available data of an experiment or model run is used. Hosting these data and making it easily accessible meets two challenges: First, the community of data users virtual organization (VO) grows and becomes more diverse, e.g. the Geographical Information System (GIS) users have started to use scientific climate data. The community investigates natural scientific as well as political, economical and social scientific questions (e.g. climate impact). And as heterogeneous as their questions are their data requirements. The second challenge lies in the continuous intensive growth of the amount of data, which is spread over different specialized data archives worldwide. Researchers create greater amounts of data in a shorter amount of time on the quickly evolving high-performance computing systems, whereas the storage amounts and network capacities develop less rapidly.

For an easy access to globally distributed climate data, a close collaboration between these data archives is necessary:

- a portal for data search and discovery, which needs an underlying uniform data description—technically by using a uniform metadata model and semantically by deriving a profile for VO purposes, e.g. including the Network Common Data Form (NetCDF) Climate and Forecast (CF) convention (cf. metadata in Chap. 4);
- triggering data access at the different data archives out of a community portal, requiring a uniform data access interface;
- solving the security problem of a single-sign-on (SSO: one authentication for authorization at all data archives) and secure data retrieval;
- (powerful) effective data selection by server-side processing gets more important to assure appropriate delivery times and to reduce the storage capacity needed at the user's home institution.

M. Stockhause (✉)
Deutsches Klimarechenzentrum, Hamburg, Germany
e-mail: stockhause@dkrz.de

W. Hiller et al., *Earth System Modelling – Volume 6*, SpringerBriefs in Earth
System Sciences, DOI: 10.1007/978-3-642-37244-5_5, © The Author(s) 2013

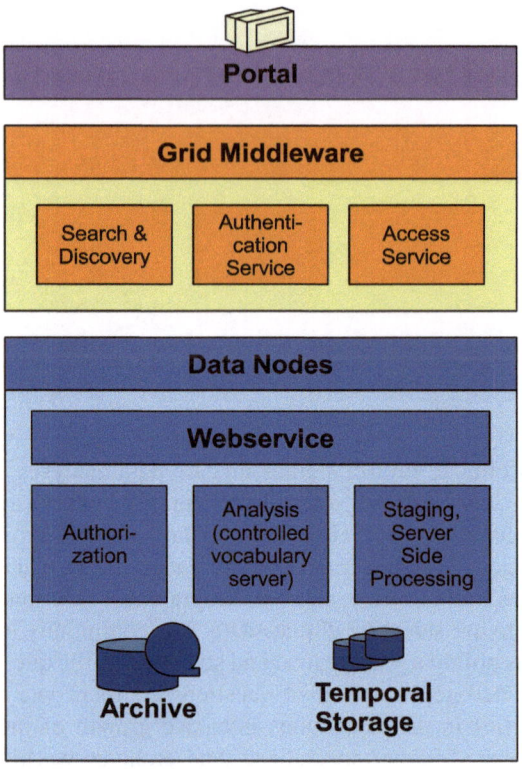

Fig. 5.1 Technical infrastructure of Data Grids

Such an infrastructure is called a Data Grid (Fig. 5.1). A prototypical architecture for homogeneous and standardized model data is currently set up as the distributed data management solution for IPCC-AR5 (Sect. 5.4). In order to serve the multiple existing vocabularies and to fulfill the requirements of the scientific community, a Data Grid with its technical infrastructure is not sufficient. Further user assistance is necessary like additional offered server-side services for data analyses and visualization and export possibilities to GIS or Google Earth (Keyhole Markup Language - KML) or a controlled vocabulary service to handle different terms used in different communities.

5.1 Existing Data Access Solutions in ESM

Nowadays local data centers have been established driven by the needs of the ESM community or a specific scientific institute. The largest are the World Data Centers (WDC), like the World Data Center for Climate (WDCC)[1] hosted at the German Climate Computing Centre (DKRZ) for ESM data. Apart from these, specific

---

[1] http://www.wdc-climate.de

data collections exist, which host usually homogeneous data, like IPCC-AR4[2] with detailed data requirements or standardizations, e.g. the self-describing NetCDF/CF data format (WGCM 2007).

Aside from these data providers a couple of national Data Grids have been developed over the last decade: Earth System Grid (ESG)[3] starting in 2001 in the USA and about 5 years later the NERC DataGrid (NDG)[4] in the UK, and the Collaborative Climate Community Data and Processing Grid (C3Grid)[5] in Germany. NDG and C3Grid are still partly a matter of research and development. The ESG is productive as data infrastructure for the Coupled Model Intercomparison Project Phase 5 (CMIP5),[6] which is to underlie the IPCC-AR5. To speed up the ESG infrastructure development the ESG Federation (ESGF, Williams et al. 2011) was founded under participation of the British Atmospheric Data Centre (BADC, a partner in NDG) and DKRZ (partner in C3Grid). The main purpose of Data Grids is data retrieval out of distributed data archives. The existing data acccess solutions in Data Grids will be described exemplarily for these three: ESG, NDG, and C3Grid.

Other kinds of grids have different foci: collaboration grids (e.g. the Geosciences Network[7]) aiming at scientific collaboration on project data and tools, and compute grids for efficient data intensive processing (e.g. Enabling Grids for E-science (EGEE)[8] and TeraGrid[9]).

For building a Data Grid out of different local data archives with their individual data standardizations, data descriptions and data access solutions several efforts are necessary to bridge the existent heterogeneity:

- uniform metadata provided by the participating data providers for data discovery,
- uniform data access interface implemented at the local data providers,
- uniform description of additional services (service metadata) available at a specific data provider.

For data discovery in a central portal as entry point, underlying homogeneous metadata descriptions (Chap. 4) are needed. Accessing data according to the user's request and independent of the data provider requires a basic agreement on access functionality. Because of the huge data amount available for an experiment or model run and the rather slow delivery via the internet, it becomes essential to offer user specific data selections. Such a basic data access functionality, which is established at most local data centers and C3Grid, commonly provide temporal, spatial, and content selections. This uniform data access interface has to be served by all data providers. As a precondition the SSO problem has to be solved.

---

[2] https://esgcet.llnl.gov:8443

[3] http://www.earthsystemgrid.org

[4] http://ndg.badc.rl.ac.uk

[5] http://www.c3grid.de

[6] http://cmip-pcmdi.llnl.gov/cmip5

[7] http://www.geongrid.org

[8] http://www.eu-egee.org

[9] http://www.teragrid.org

Some data providers offer additional services, which reach from rather simple format conversions up to complex data analyses, transformations and statistical calculations for a single or for multiple files, or visualization opportunities. The importance of powerful server-side access and analysis services increase due to the rapid growth in data amount spread over multiple data centers (e.g.Williams et al. 2009). Among these visualization and derived climate indices and extremes become more important for an easy access of scientific data for climate impact research and for non-scientists (Frich et al. 2002). Furthermore for the increasingly complex questions in climate impact research, intensive interdisciplinary collaborations are demanded between scientists of different professions as well as decision makers in politics, which introduces new questions into the VO. Therefore in addition to the "power" users, researchers working with cutting-edge models, an at least one magnitude larger number of analysts from different communities work on or with ESM data (Williams et al. 2008a).

The challenge is how to make these local services globally available. The establishment of a global network out of smaller regional ones (similar to e.g. the Global Telecommunication System (GTS) of the World Meteorological Organization (WMO); Hankin et al. 2009) and the description of the offered services by metadata are required. International standards for these service metadata (e.g. ISO 19119) should be used, desirably with further conventions, like name conventions comparable with the NetCDF/CF standard for data descriptions. Tools and services are offered in ESG II[10] (Chap. 7, and Williams et al. 2009) and C3Grid (see Chap. 6), which still lacks a uniform description. Therefore their applications require special user knowledge or a preselection of suitable data in the central portal. For transparency and an easy deployment of the services, the deployment of software packages using established referenced tools, like unidata tools,[11] is recommendable.

A second aspect of the increased VO diversity is the introduction of heterogeneous vocabularies. Therefore an additional central vocabulary service (also called service translation library) with underlying descriptions of the terms and their complex relationships are necessary (ontology). This enables multipurpose data access by knowledge sharing and reuse (Pouchard et al. 2005). Therefore ontology development provides a general approach for an interoperation between different communities as well as for different data access gateways with different metadata models (e.g. ESG and NDG developed a shared ontology for collaboration; Pouchard et al. 2003). Instead of a simple mapping functionality appropriate for unique relationships between metadata entries of different metadata models, ontologies can represent complex relationships. Ontologies demand continuous development efforts to be kept up-to-date.

---

[10] http://www.earthsystemgrid.org

[11] http://www.unidata.ucar.edu/software

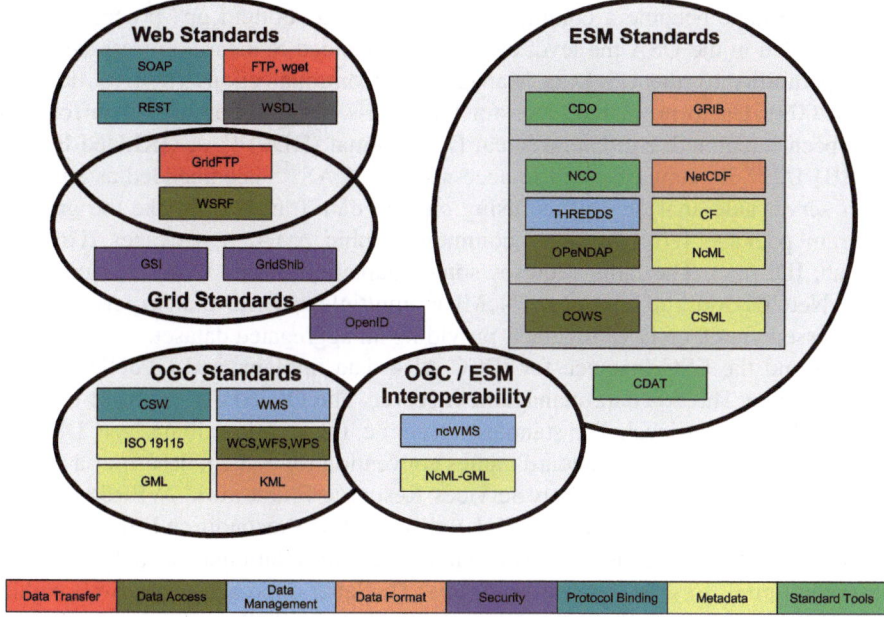

Fig. 5.2 Technical and community standards used in ESM Data Grids.

5.2 Technical Solutions for Data Access

For data access different communities have independently developed technical solutions. Within the climate community several data access solutions have been established. Some are widely used like the Open-source Project for a Network Data Access Protocol (OPeNDAP)[12] interface. Apart from the climate community the GIS community Open Geospatial Consortium (OGC) as well as the World Wide Web Consortium (W3C)[13] and grid communities like the Open Grid Forum (OGF),[14] the Globus alliance[15] or the Organization for the Advancement of Structured Information Standards (OASIS)[16] have established standards for web based data retrieval (Fig. 5.2).

The Open-source Project for a Network Data Access Protocol (OPeNDAP) started as a specific development in oceanography for NetCDF flat file access. It is usually implemented together with an underlying Thematic Realtime Environmental Distributed Data Services (Thredds)[17] data management service. NetCDF, Thredds and

[12] http://www.opendap.org

[13] http://www.w3c.org

[14] http://www.ogf.org

[15] http://www.globus.org

[16] http://www.oasis-open.org

[17] http://www.unidata.ucar.edu/projects/THREDDS

OPeNDAP have become a community standard in a "bottom up" process, which has achieved in the USA the level of a "Recommended Standard" for gridded data interoperability by the US Data Management Communications (DMAC; Hankin et al. 2009). Up to now, interfaces for the access of other common data formats have been integrated, e.g. Hierarchical Data Format (HDF) [18] or GRIdded Binary (GRIB) 1/2.[19] In some cases a life access server (LAS)[20] is embedded as a higher order server-side analysis service using climate data frameworks like the graphic program package ferret or other common graphic program packages (GrADS, Matlab, IDL,...). The latter requires some adaptation effort. With additional use of the NetCDF Markup Language (NcML)[21] multiple datasets of an experiment can be accessed via OPeNDAP together, providing an aggregated dataset.

ESG and the ESG instance for IPCC-AR4 data implemented OPeNDAP with an underlying Thredds data management server. The OPeNDAP protocol directly triggers the data retrieval. The standards used, i.e. the self-describing NetCDF data format with the use of CF standard names convention for content description makes an interface definition (like Web Services Resource Framework (WSRF) or Web Services Description Language (WSDL)) dispensable. Precondition is therefore the acceptance of the standards used in the whole VO or an additional vocabulary service for communities to serve other user groups. The latter approach is followed within ESG II (Williams et al. 2009) and NDG (Latham et al. 2009).

The OGC developed the Web Coverage Service (WCS; Whiteside and Evans 2008) and the Web Feature Service (WFS) as standards for data retrieval, including a subset functionality. They support the common 2-dimensional raster (only geo-rectified grids) or vector (point, line/polyline, polygon) data in GIS applications. Vertical and temporal information definitions are not sufficient, because vertically only length-based coordinates and only a single temporal information are supported (Nativi et al. 2006).

NDG decided to move towards these OGC services. A python based web framework (Centre of Environmental Data Archival (CEDA) OGC Web Services (COWS), Lowe et al. 2009; Stephens et al. 2012) is developed on top of the Climate Science Modeling Language (CSML) API (Lawrence et al. 2009). CSML is based on the OGC/ISO standard for the description of geographical data the Geography Markup Language (GML), but has several extensions, e.g. in grid description, and includes an additional access section. The NDG developed a data access through Website Management Framework (WMF)/WCS and Web Map Service (WMS) protocols, the NetCDF Web Map Service (ncWMS),[22] which was integrated into Thredds 4.0 as a new individual data access interface. A couple of projects dealing with observation data have already implemented this Thredds version, e.g. the Integrated Marine Observing System (IMOS; Proctor et al. 2009) or the Marine Environment and

---

[18] http://www.hdfgroup.org

[19] http://www.wmo.int/pages/prog/www/DPS/FM92-GRIB2-11-2003.pdf

[20] http://ferret.pmel.noaa.gov/LAS

[21] http://www.unidata.ucar.edu/software/netcdf/ncml

[22] http://www.resc.rdg.ac.uk/trac/ncWMS

Security for the European Area (MERSEA) project (GODIVIA2 2009; Blower et al. 2009). These portals utilize OpenLayers[23] as OGC WMS-client, which provides an export as OGC KML, used e.g. by Google Earth. As alternative to COWS, NDG implemented for interoperability an OPeNDAP interface based on the pure Python library implementation of the OPeNDAP protocol (Pydap) for interoperability. COWS and Pydap data access currently run in the NDG development discovery interface.[24] Additionally, a data processing via the OGC WPS protocol with local result caching was developed within COWS.

C3Grid developed and implemented an interface for data access based on W3C/ Globus Tool Kit (GTK, a widely used grid software package) standard WSRF, which has a subsetting functionality in respect to spatial-temporal and content selection (Kindermann et al. 2007; Plantikow et al. 2009). The WSRF has been replaced recently by a RESTful web service (see below).

The OGC Geo-interface for Atmosphere, Land, Earth, and Ocean NetCDF (GALEON) project [25] has a more general approach in making community Thredds/ OPeNDAP services OGC-compliant. For data access not only WCS and WMS interfaces to the Thredds server but an integration of WCS into the OPeNDAP protocol using NcML-GML (GML enabled NcML; Nativi et al. 2004) are developed. Within the OGC the Meteorology and Oceanography Domain Working Group (MetOceanDWG) was formed in 2009, which enables collaboration between the OGC and ESM communities. The MetOceanDWG is co-chaired by the WMO and focusses on the development of community profiles of common standards and derive best practises for the implementation of technical standards. MetOceanDWG started with voluntary interoperability tests. WMO and OGC signed a formal Memorandum of Understanding in November 2009, which backs this collaboration (WMO and OGC 2009).

Other harmonization efforts are undertaken to ease data access out of different archives of geographical data, like the European Infrastructure for Spatial Information in Europe initiative (INSPIRE The European Parliament and Council 2007), which recommends data discovery and access by OGC services as a default service technology (INSPIRE drafting team "Network Services" 2008b; INSPIRE 2009): OGC WFS (Web Feature Service; ISO 19142)/WCS and FE (Filter Encoding; ISO 19143) for spatial-temporal selection. With GIS users becoming a part of the climate VO— as data users as well as data providers— such interoperabilities between OGC and ESM data access standards increase in importance. The second generation Data Grids will presumably support other communities with other vocabularies and other standards, e.g. the OGC standards. Figure 5.3 gives an overview over the data access solutions used for ESM data, reaching from low level services for all kinds of data like FTP to high level inter-community services (e.g. server-side analyses or grid workflows), which require detailed metadata descriptions.

---

[23] http://openlayers.org

[24] http://ndgbeta.badc.rl.ac.uk

[25] http://www.ogcnetwork.net/galeon

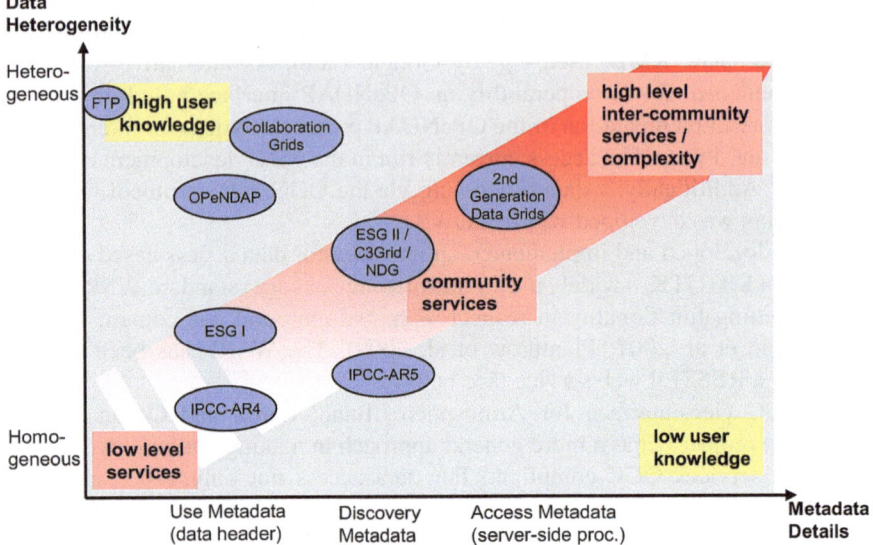

Fig. 5.3 Community data access solutions regarding to data heterogeneity and metadata detail. Where heterogeneous data without metadata additional to the data header information requires specialist user knowledge for interpretation, well described data could serve less aknowledged users. If the data is heterogeneous, complex services for this inter-community data access are essential, possibly with the use of a vocabulary service

In addition to a data access interface a protocol binding is essential, through which information is exchanged between the local data provider and the Data Grid middleware. Representational State Transfer (REST, e.g. Fielding and Taylor 2002) is a (technically) simple protocol, a HTTP request method for a command-line based data access and without any messaging layer in a distributed system. Therefore, e.g. asking for the status of the user request or canceling it during execution is not possible using REST protocols. Such a messaging layer is present in a web service consisting of the Simple Object Access Protocol (SOAP) (Gudgin et al. 2007) and a webservice interface definition, i.e. a machine-processable description of service (for a comparison of REST and SOAP cf. Haas 2005). SOAP is increasingly used as protocol for data accesses and recommended, e.g. by INSPIRE.

The security problem to solve is an authorization without user interaction (client-based data accesses) by an automated (web-based) data access mechanism. For Compute Grids and open source data in Data Grids, the certificate solution for user identification is usually sufficient. Every user with a certificate signed by a trusted VO is accepted and grants him compute time or the access of the open source data, respectively. To prevent an extended misuse of certificates only short-living so called proxy-certificates are sent for request delegation. This mechanism is called Grid

Security Infrastructure (GSI)[26] and is part of the Globus Tool Kit. However, dealing with data with restricted access, more information about user permissions is required for authorization. This information has to be passed to the local data provider together with the certificate. Moreover the participating data providers have to agree on assertions (e.g. as Security Assertion Markup Language (SAML) callouts), which grant additional permissions for the retrieval of special data aggregations, and to trust the authentication authority of another member of the VO.

One research approach is GridShib,[27] a combination of GSI and Shibboleth,[28] a federated attribute-based authorization mechanism, which is in a prototypical state. For the federated data management solution of IPCC-AR5 / ESG II, an approach using OpenID[29] for authentication and additional attribute services for authorization is in use. In contrast to Shibboleth, OpenID has no administrative rules about building a federation of organizations and trusting attributes (assertions), which therefore has to be figured out by the VO itself (Sect. 5.4). The three Data Grids under consideration follow a dual approach: establishing a short-term pragmatic solution including OpenID, especially for IPCC-AR5/CMIP5, and following a long-term Shibboleth-enabled solution: C3Grid (Groeper et al. 2009) and NDG (Lawrence et al. 2007).

The transfer from the local data provider to the user via the internet is performed by gridFTP, FTP, wget or HTTP. Though the gridFTP access is both save and highly performant, it has only been used for internal transfers among the data providers, but not for data download in the existing productive Data Grids, so far. ESG II (Williams et al. 2008b, 2009) and IPCC-AR5 intends to use a client-based access based on https access and a simple delivery approach, e.g. by wget. C3Grid uses FTP.

5.3 Aspects of Local Implementations

Becoming a data provider in a Data Grid is associated with efforts:

- implementation of the community data access interface,
- provision of server-side analyses to serve the interface and provide additional services, and
- implementation of the VO security approach.

For the data request a web service has to be set up, which accepts the user requests and triggers the data retrievals. At first the user is identified at the local data provider by the Distinguished Name (DN) in the proxy certificate or another unique VO identifier, and authorized by the help of additional assertions (Sect. 5.2). The data

---

[26] http://www.globus.org/security/overview.html

[27] integrating federated authorization infrastructure (Shibboleth) with Grid technology (the Globus Toolkit) to provide attribute-based authorization for distributed scientific communities, http://gridshib.globus.org.

[28] http://shibboleth.internet2.edu

[29] http://openid.net

provider has to trust the authentication and implement a mechanism for automatic user authorization based on the analysis of the additional (SAML) assertions in the user certificate. Since the data provider hosts data owned by others, he is responsible for the fulfillment of the given access and use constraints defined by the owner within the grid authorization process. Among the existing Data Grids SAML assertions are used, but for the authorization mechanism different approaches are followed by ESG II (Siebenlist et al. 2009), NDG (Lawrence et al. 2007), and C3Grid (Groeper et al. 2009), since this is still a matter of research. Generally, OpenID is judged as a ready-to-use short term solution with the long-term perspective of an implementation of Shibboleth.

The data access itself consists of a request analysis, the staging of the base datasets and the server-side processing (selection and merging) to provide the requested data. During the analysis additional local (use) metadata is retrieved from the local data management system, which orchestrates the data retrieval process. Use metadata is either stored in the file header (e.g. in NetCDF) or provided separately at the data provider. For data on tapes, a storage of use metadata on disk might be necessary for self-describing formats as well to avoid data stagings during data discovery. This has to be integrated into the local data management system.

In some cases where user request (discovery) metadata differs from local (use) metadata, an interpretation of the request has to be performed before, e.g. at C3Grid data providers; e.g. for some specialized content parameters of model outputs, standardized and VO-wide well-known names (CF) are unavailable or other standards are already established at the data providers and their institutes, like model variable names or GRIB code numbers. Controlled vocabulary servers can be used to perform the mapping of VO names to local names, running as local or as central grid service. For OPeNDAP (and LAS) based data access such an interpretation step is obsolete, because user's data selection is done on use metadata, extracted from the NetCDF headers with an aggregation possibility using NcML. When the data access procedure is orchestrated, it commonly runs in a private scratch directory. The result is delivered in a Data Grid accessible workspace. For simple requests without server-side processing, the dataset can be directly parsed out of the data archive to the user as data stream.

The main problems to be solved for server-side processing scalability are: the allocation of storage during processing, and the control of net traffic for performing multiple parallel user requests. To minimize the implementation effort and in order to guarantee the equality of server-side processings, it is desireable to use the same set of tools known and accepted in the scientific community, like the Unidata Climate Data Analysis Tools (CDAT)[30] or NetCDF Operator (NCO)[31] for NetCDF and Climate Data Operators (CDO)[32] for GRIB data formats. Several projects aim to make these tools accessible via a web service for individual, flexible server-side data process-

---

[30] http://www2-pcmdi.llnl.gov/cdat

[31] http://nco.sourceforge.net

[32] http://www.mpimet.mpg.de/cdo

ing, e.g. the Infrastructure for the European Network for Earth System Modelling (IS-ENES) and Climate Analytics on Distributed Exascale Data Archives (ExArch).

Becoming a data provider in a Data Grid might imply considerable efforts, since the Data Grid interface implementation and maintenance have to be performed in addition to the established local data access interface.

5.4 Future Developments

Nowadays most of the productive data portals are local data collections with a web interface (e.g. the ESG IPCC-AR4). For the data retrieval of CMIP5/IPCC-AR5 data the expected amount is about 10–100 times larger than for CMIP3/IPCC-AR4, where over 35 TeraByte (TB) of model data were stored (Williams et al. 2008a). An experiment in CMIP3 produced about 9 TB/model, whereas for CMIP5 about 100 TB/model are expected (Lautenschlager 2008). The resulting increase of a factor of about 10 is due to higher model resolutions, additional ESM components with additional raw data output considered, and a higher number of ensemble realizations performed for each projection. This leads to estimations for the total stored data for IPCC-AR5 between more than 2 PetaByte (PB) to several PBs including additional data by Williams et al. (2008a) and about 3 PB by Lautenschlager (2008), i.e. a higher number of individual files have to be managed by the data nodes. For the total data volume increase of CMIP5 compared to CMIP3 a factor of up to 100 is expected (Taylor et al. 2011).

The download rate is expected to grow not only due to the increased data volume but as well due to more users, causing increased server work loads. For IPCC-AR4 a peak of 900 GigaByte/month was observed around March 2007 (Williams et al. 2009). Therefore the higher order server-side services OPeNDAP and LAS are set up in order to reduce user's data download volume.

Thus, a federated approach is developed for IPCC-AR5 data access, serving data distributed over several geographically separated storage locations (Fig. 5.4): a Data Grid. Detailed prescriptions (Taylor 2009; Taylor et al. 2011) for the internal data structure (Taylor and Doutriaux 2011) as well as naming and directory structure conventions (Taylor et al. 2009) exist. The ESGF sets up a data infrastructure as an instance of the ESG II infrastructure (Williams et al. 2008b).

Data nodes were set up at nearly all national data centers or participating modeling centers, which publish the CMIP5 data to the ESG including data versioning. Accompanied metadata initiatives, Common Metadata for Climate Modelling Digital Repositories (METAFOR; Guilyardi et al. 2011) and Earth System Curator (Dunlap et al. 2008), built a Common Information Model (CIM) for a detailed description of ESMs. The developed uniform complete metadata model for the description of data, data access, model, experiment and simulation (model application) is used within CMIP5/IPCC-AR5.

The ESGF gateways harvest the metadata from the data nodes (Thredds catalogs) and the CIM repository for data search and discovery functionalities. Currently,

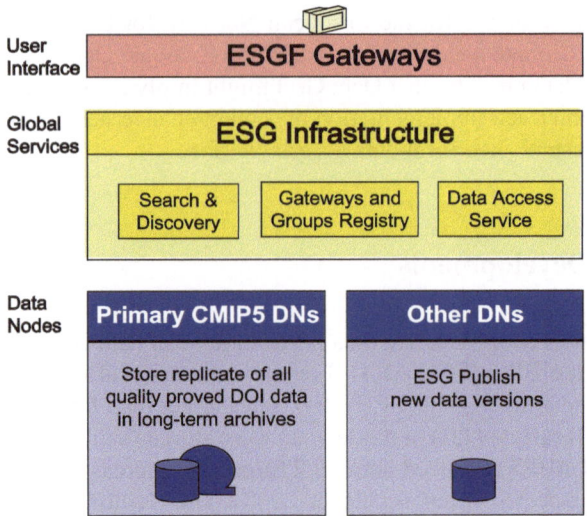

Fig. 5.4 ESG infrastructure for the federated data access of the CMIP5/IPCC-AR5 data based on the ESG II three tiers infrastructure (*DN* data node)

ESGF gateways are located in the USA, Europe, and Australia. Among these gateways with data nodes the three primary CMIP5 data centers PCMDI/LLNL, BADC and WDCC/DKRZ committed themselves to long-term archive the most relevant part of the CMIP5 data (data used for IPCC-AR5), especially those data versions with assigned DataCite[33] Digital Object Identifier (DOI), i.e. citation regulations. Therefore they hold data replica of about 10–20 % of the provided CMIP5 data (marked as *output1*) available. The national Data Grids NDG and C3Grid plan to provide access to CMIP5 data via their portals as well.

The overall security concept is based on SSO with OpenID (see Sect. 5.2) and a web-based data access, involving https protocol and SAML callouts for receiving the authorization assertions. Data is to be easily-accessible, e.g. by wget and multiple data download scripts using wget.

The ESGF infrastructure development has the potential to become the core of the international Data Grid software development in Earth System Sciences. It is a joint development of the grid infrastructure developers in the USA and Europe. Other grid developers, who are currently software users of ESGF, could become software developers. Thus it is expected that the data grid development will concentrate along the ESGF infrastructure development providing the core services. National and European grid initiatives as well as observation data grids can make use of the ESGF core services and develop specific own services for the special needs of their user community, like server-side processing services.

For semantic interoperability between different initiatives vocabulary services will increase in importance. Along with the data and data-central information, model

---

[33] http://datacite.org

and application information as well as tools and services will be shared for data interpretation and analysis.

5.5 Conclusions

- The ESGF initiative bundles the international data grid development efforts to provide software and tools for core services for Earth System Science (ESS) Data Grids. Data Grids with their geographically distributed data storage are essential to fulfill future download rates.
- Sophisticated and powerful server-side services, e.g. effective data selection, vizualisation, and derived climate indices, are essential to deal with the expected high data volumes and the expected diversity of user requests. Moreover server-side processing will provide more flexible user-defined accesses to ESM data.
- Additional services like a controlled vocabulary server enables different user communities to access and analyse climate data.
- The security problem of a single-sign-on is still a topic of research, technically (web-based data access) as well as administratively (rules for building a VO).
- Metadata schema developments and standardizations in the closely co-operating initiatives Earth System Curator and METAFOR support uniform and detailed additional information on data, enabling faceted data search in portals. Data provenance is significantly improved from pure data central descriptions toward an overall description of data history. Moreover the additional descriptions of models and tools and their application become connected to the (self-describing) data.
- In the technical approaches for data access standards, a gap between the GIS world of OGC standards and the ESM world is present, which hinders interoperability. For the harmonizations in geographic data access OGC standards are recommended, e.g. in the European INSPIRE guideline. Inside the OGC initiatives have been established for the interoperability between ESS and OGC: GALEON aims at making ESM tools (e.g. OPeNDAP and Thredds) OGC compliant, whereas in the MetOceanDWG the WMO interacts with OGC in the further development of OGC services. It is expected that second generation Data Grids in ESS will provide apart from the traditional established interfaces an access via OGC services.

References

Blower J, Haines K, Santokhee A, Liu C (2009) GODIVA2: interactive visualization of environmental data on the Web. Pjil Trans R Soc A 367:1035–1039. doi:10.1098/rsta.2008.0180

Dunlap R, Mark L, Rugaber S, Balaji V, Chastang J, Cinquini L, DeLuca C, Middleton D, Murphy S (2008) Earth system curator: metadata infrastructure for climate modelling. Earth Sci Inform 1:131–149. doi:10.1007/s12145-008-0016-1

Fielding RT, Taylor RN (2002) Principled design of the modern Web architecture. ACM Trans Internet Technol 2(2):115–150. doi:10.1145/514183.514185

Frich P, Alexander L, Della-Marta P, Gleason B, Haylock M, Tank AK, Peterson T (2002) Observed coherent changes in climatic extremes during the second half of the twentieth century. Clim Res 19:193–212. http://www.vsamp.com/resume/publications/Frich_et_al.pdf

GODIVIA2 (2009) User guide of the GODIVA2 web client, http://www.resc.rdg.ac.uk/trac/ncWMS/wiki/GodivaTwoUserGuide

Groeper R, Grimm C, Makedanz S, Pfeiffenberger H, Ziegler W, Gietz P, Schiffers M (2009) A concept for attribute-based authorization on D-Grid resources. Future Gener Comput Syst 25(3):275–280. doi:10.1016/j.future.2008.05.008

Gudgin M, Hadley M, Mendelsohn N, Moreau JJ, Nielsen HF, Karmarkar A, Lafon Y (eds) (2007) SOAP version 1.2 Part 1: messaging framework (2nd edn). Project website, World Wide Web Consortium (W3C). http://www.w3.org/TR/soap12

Guilyardi E, Balaji V, Callaghan S, DeLuca C, Devine G, Denvil S, Ford R, Pascoe C, Lautenschlager M, Lawrence B, Steenman-Clark L, Valcke S (2011) The CMIP5 model and simulation documentation: a new standard for climate modelling metadata. CLIVAR Exch 16(56):42–46

Haas H (2005) Reconciling Web services and REST services. In: proceedings of the 3rd IEEE European conference on Web services (ECOWS) 2005, Växjö, Sweden, 14–16 Nov 2005. http://www.w3.org/2005/Talks/1115-hh-k-ecows

Hankin S, Blower JD, Carval T, Casey KS, Donlon C, Lauret O, Loubrieu T, de la Villeon LP, Srinivasan A, Trinanes J, Godøy Ø, Mendelssohn R, Signell R, de La Beaujardiere J, Cornillon P, Blanc F, Rew R, Harlan J (2009) NetCDF-CF-OPeNDAP: standards for ocean data interoperability and object lessons for community data standards processes (Draft). In: Proceedings of OceanObs'09, 21–25 Sept 2009

INSPIRE (2009) Draft technical guidance for INSPIRE download services—Version 2.0, 25 Sept 2009. http://is.gd/894qi

INSPIRE drafting team "Network Services" (2008b) INSPIRE network services architecture—version 3.0 19 July 2008. http://is.gd/8949v

Kindermann S, Stockhause M, Ronneberger K (2007) Intelligent data networking for the earth system science community. In: Proceedings from the German e-Science conference 2007, Max Planck Digital Library, Baden-Baden, Germany, ID: 316512.0. http://edoc.mpg.de/316512

Latham S, Cramer R, Grant M, Kershaw P, Lawrence B, Lowry R, Lowe D, O'Neill K, Miller P, Pascoe S, Pritchard M, Snaith H, Woolf A (2009) The NERC DataGrid services. Philos Trans Royal Soc A Math Phys Eng Sci 367(1890):1015–1019. doi:10.1098/rsta.2008.0238

Lautenschlager M (2008) IPCC TGICA and IPCC DDC for AR5 Data. In: Proceedings of the GO-ESSP meeting, Seattle, 17–19 Sep 2008. http://go-essp.gfdl.noaa.gov/2008/files/lautenschlager_GO-ESSP-TGICA+AR5-Seattle-17-19090.ppt

Lawrence BN, Kerschaw P, Blower J (2007) Practical access control using NDG security. In: Proceedings of the UK e-science all hands meeting 2007, 10–13 Sept 2007. http://www.allhands.org.uk/2007/proceedings/papers/788.pdf

Lawrence B, Lowry R, Miller P, Snaith H, Woolf A (2009) Information in environmental data grids. Philos Trans Royal Soc A Math Phys Eng Sci 367(1890):1003–1014. doi:10.1098/rsta.2008.0238

Lowe D, Woolf A, Lawrence B, Pascoe S (2009) Integrating Integrating the climate science modelling language with geospatial software and services. Int J Digit Earth 2(1):29–39. doi:10.1080/17538940902866161

Nativi S, Caron J, Domenico B (2004) NcML-GML: encoding NetCDF datasets using GML. In: Proceedings of the 15th international workshop on database and expert systems applications, 30 Aug–3 Sept 2004, pp 804–808

Nativi S, Domenico B, Caron J, Davis E, Bigagli L (2006) Extending THREDDS middleware to serve OGC community. Adv Geosci 8:57–62. http://www.adv-geosci.net/8/57/2006

Plantikow S, Peter K, Högqvist M, Grimme C, Papaspyrou A (2009) Generalizing the data management of three community grids. Future Gener Comput Syst 25(3):281–289. doi:10.1016/j.future.2008.05.001

Pouchard L, Woolf A, Bernholdt DE (2005) Data grid discovery and semantic web technologies for the earth sciences. Int J on Digit Libr 5(2):72–83

Pouchard L, Cinquini L, Drach B, Middleton D, Bernholdt D, Chanchio K, Foster I, Nefedova V, Brown D, Fox P, Garcia J, Strand G, Williams D, Chervenak A, Kesselman C, Shoshani A, Sim A (2003) An ontology for scientific information in a grid environment: the Earth System Grid. In: Proceedings of the 3rd IEEE/ACM international symposium on cluster computing and the grid (CCGRID'03). http://ieeexplore.ieee.org/stamp/stamp.jsp?arnumber=01199424

Proctor R, Roberts K, Bohm P, Cameron S, Hope J, Jones C, Mancini S, Pepper K, Tattersall K, Ward B, Williams G, Mak P, Goessmann F (2009) Information infrastructure for the Australian integrated marine observing system. In: Proceedings of OceanObs'09, 21–25 Sept 2009

Siebenlist F, Ananthakrishnan R, Bernholdt D, Cinquini L, Foster I, Middleton D, Miller N, Williams D (2009) Earth system grid authentication infrastructure: integrating local authentication, openID and PKI. In: Proceedings of the TeraGrid 09, Virginia, 22–25 June 2009. http://www.teragrid.org/tg09/files/tg09_submission_79.pdf

Stephens A, James P, Alderson D, Pascoe S, Abele S, Iwi A, Chiu P (2012) The challenges of developing an open source, standards-based technology stack to deliver the latest UK climate projections. Int J Digit Earth 5(1):43–62. doi:10(1080/17538947),2011,571724

Taylor K (2009) CMOR dimensions. Technical report. http://cmip-pcmdi.llnl.gov/cmip5/output_req.html

Taylor KE, Balaji V, Hankin S, Juckes M, Lawrence B (2009) CMIP5 and AR5 data reference syntax (DRS). Technical report. http://is.gd/895GZ

Taylor K, Doutriaux C (2011) CMIP5 model output requirements: file contents and format, data structure and metadata. Technical report. http://pcmdi-cmip.llnl.gov/cmip5/docs/CMIP5_output_metadata_requirements.pdf

Taylor K, Stouffer R, Meehl G (2011) An overview of CMIP5 and the experiment design.Submitted to Bulletin of the American Meteorological Society. doi:10.1175/BAMS-D-11-00094.1

The European Parliament and Council (2007) Directive 2007/2/EC of the European Parliament and of the council of 14 March 2007 establishing an Infrastructure for Spatial Information in the European Community (INSPIRE). Off J Eur Union 50:1–14. http://eur-lex.europa.eu/JOHtml.do?uri=OJ:L:2007:108:SOM:EN:HTML

WGCM (2007) IPCC standard output from coupled ccean-atmosphere GCMs. Technical report. http://www-pcmdi.llnl.gov/ipcc/standard_output.html

Whiteside A, Evans JD (2008) Web coverage service (WCS)—implementation standard. Project website, open geospatial consortium (OGC). http://www.opengeospatial.org/standards/wcs

Williams D, Ananthakrishnan R, Bernholdt D, Bharathi S, Brown D, Chen M, Chervenak A, Cinquini L, Drach R, Foster I, Fox P, Fraser D, Garcia J, Hankin S, Jones P, Middleton D, Schwidder J, Schweitzer R, Schuler R, Shoshani A, Siebenlist F, Sim A, Strand W, Su M, Wilhelmi N (2009) The Earth System Grid: enabling access to multimodel climate simulation data. Bull Amer Meteor Soc 90:195–205. doi:10.1175/2008BAMS2459.1

Williams D, Ananthakrishnan R, Bernholdt D, Bharathi S, Brown D, Chen M, Chervenak A, Cinquini L, Drach R, Foster I, Fox P, Hankin S, Henson V, PJones, Middleton D, Schwidder J, Schweitzer R, Schuler R, Shoshani A, Siebenlist F, Sim A, Strand W, Wilhelmi N, Su M (2008a) Data management and analysis for the Earth Sytem Grid. J Phys 125. doi:10.1088/1742-6596

Williams D, Middleton D, Anitescu M, Balaji V, Bethel W, Cotter S, Strand G, Schuchardt K, Shoshani A (2008b) Extreme scale data management, analysis, visualization, and productivity in climate change science, panel report. In: Extreme scale computing workshop, Dec 2008. http://is.gd/8950r

Williams D, Taylor K, Cinquini L, Evans B, Kawamiya M, Lautenschlager M, Lawrence B, Middleton D, ESGF contributors (2011) The Earth System Grid Federation: software supporting CMIP5 data analysis and dissemination. CLIVAR Exch 16(56):40–42

WMO, Ogc (2009) Memorandum of understanding between the World Meteorological Organization and the Open Geospatial Consortium, Inc. In: Proceedings of the 2nd workshop on the use of GIS/OGC standards in meteorology, Toulouse, 23–25 Nov 2009. URL http://is.gd/896Ab

Chapter 6
Collaborative Climate Community Data and Processing Grid—C3Grid: Workflows for Data Selection, Pre- and Post-Processing in a Distributed Environment

Bernadette Fritzsch and Wolfgang Hiller

Grid technology can help scientists to overcome problems with the Data Deluge in climate research, as it facilitates large scale data sharing and reuse of data. Worldwide there are different initiatives to facilitate data handling for scientific work and to built up collaborational working environments with interinstitutional data access. One example is the Collaborative Climate Community Data and Processing Grid (C3Grid) which[1] aims at building up a grid infrastructure for a seamless and fast access to the commonly used data resources of German data archives. In the initial project phase, C3Grid was one of five academic community projects within the German D-Grid Initiative.[2] Between 2005 and 2009 a prototype grid based infrastructure was developed and implemented. In the current second project phase the collaborative research environment is advanced for broader user requirements with special focus on workflows and with respect to international interoperability. Both projects have been funded by the German Ministry of Education and Research (BMBF).

6.1 General Remarks

Climate research and Earth System sciences strongly depend on different data sources from simulation runs of coupled Earth system models and a multitude of observational data. These data sets are distributed over many archives and sites worldwide and have different states with respect to general accessibility and data quality. The

---

[1] http://www.c3grid.de
[2] http://www.d-grid.de

B. Fritzsch (✉) · W. Hiller
Alfred-Wegener-Institut für Polar- und Meeresforschung,
Am Handelshafen 12, 27570 Bremerhaven, Germany
e-mail: bernadette.fritzsch@awi.de

individual researcher is not only drowned by the Data Deluge but in his daily work also faces obstacles originating from data diversity. Therefore progress in scientific work can only be achieved by connecting several data sources and taking into account all available information and metadata related to the individual data sets. In reality this is rather complicated, not only by distributed archiving of data through different institutions but also by the heterogeneity in the data description and differences in data access modes.

It is obvious, that only a common platform with uniform access mechanisms and standardized data descriptions can substantially improve the scientists' ability to use these data in a successful way to explore and explain the changes in the Earth system.

The C3Grid mission statement essentially reflects this situation by

- enhancing uniform and easy access to distributed and heterogeneous data sources,
- adopting a workflow oriented approach,
- provisioning standardized data descriptions based on ISO standards, and
- delivering combined storage and processing facilities for scientists.

Meeting these requirements, several major German Earth science institutions have teamed up with partners from information sciences to pool resources for developing a community grid infrastructure. The consortium includes different players: data providers as well as workflow developer and scientific user (Fig. 6.1). Within the data providers are the world data centers for Climate (WDC-Climate), for Remote

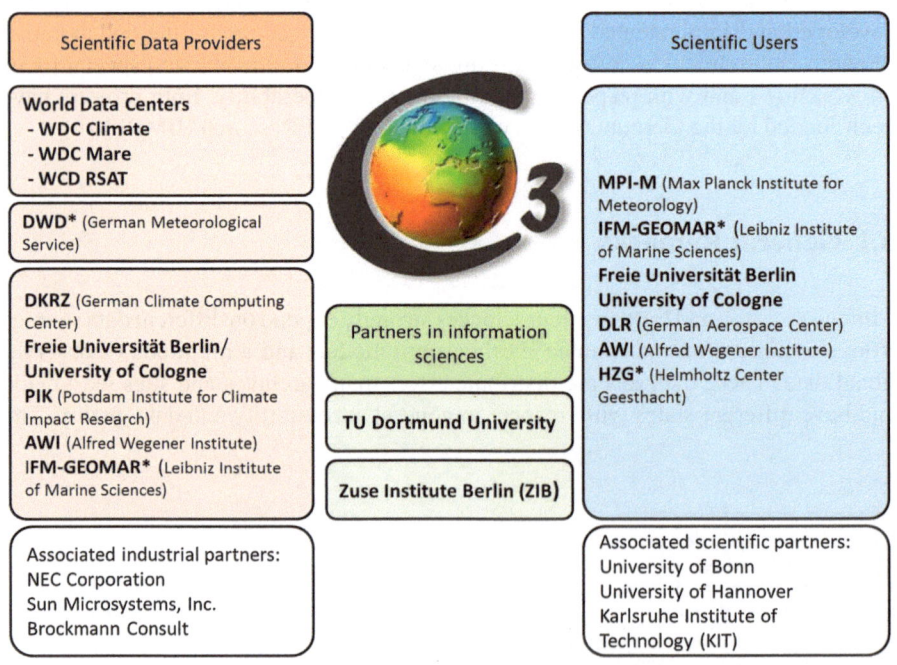

Fig. 6.1 Consortium of C3Grid (An *asterisk* * marks partners in only first project phase)

Sensing of the Atmosphere (WDC-RSAT) and for Marine Environmental Sciences (WDC-Mare) as well as Germany's National Meteorological Service (DWD). Furthermore, partners with grid experience from Zuse Institute Berlin (ZIB) and University Dortmund develop community specific components in C3Grid.

6.2 Architecture and Middleware Components

The C3Grid architecture (see Fig. 6.2) reflects the requirements of the potential users. It offers a service-oriented architecture that realizes distributed data access, transport and analysis on several layers of abstraction. The resource level integrates data providers and computational resources to a virtual workspace which is used by higher-level Grid services to execute complex workflows. On the highest level, the web-based portal application provides a simple-to-use general entry point for users. The first version of C3Grid was based on Globus Toolkit 4.x augmented by some C3Grid specific components. In the meantime a technological shift occured and Globus Toolkit 5 is used. Central components will be described in the following section.

The overall strategic decision for the C3Grid architecture was to adhere to standards and the Global Grid Forum (GGF) or D-Grid supported software components whenever possible. Only in areas where no substantial software offerings or packages

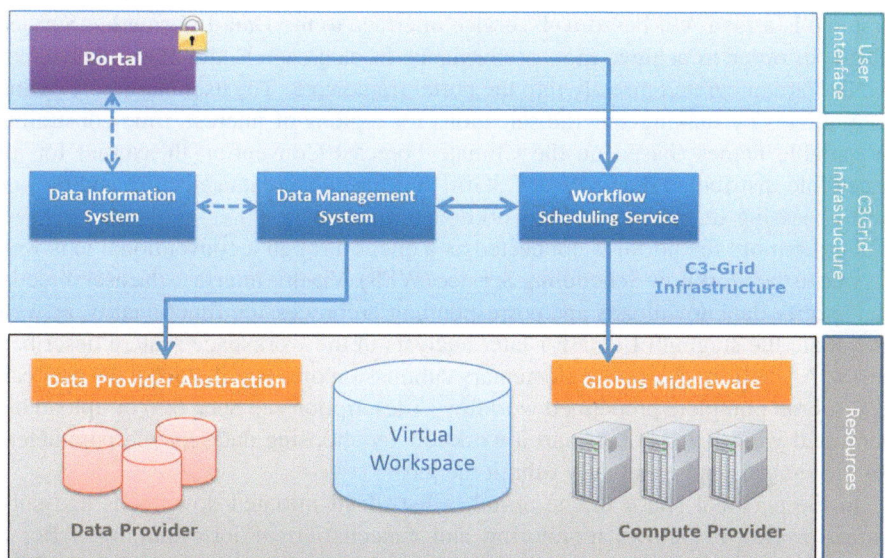

Fig. 6.2 General architecture of C3Grid

were on the market, C3specific software was implemented. Still there was general agreement to adhere to common definitions and standards whenever possible.

6.2.1 C3Grid Portal

The portal is the central focus point for user interaction with the C3Grid resources. It hides all technical details of data and compute providers as well as services from the infrastructure layer from the user's perspective. Using the central metadata catalogue (Data Information System) the portal allows data browsing, selection, and subsequent download of data. This process is supported by a download assistant which triggers so called preprocessing operations at C3Grid data provider sites and includes mechanisms for cutting out geographical regions, time slices, etc. in order to reduce the amount of data to be transferred and transparently provide requested data subsets to the user. The current implementation of the C3Grid portal is based on a Java Specification Request (JSR) - 268 compliant open source portal framework, the Liferay enterprise portal framework.[3] Future releases of the portal will contain a switch from a "simple" framework to a full Content Management System to handle the requirement of multiple portals for specific user groups like scientists, interested public, insurance companies, decision makers and other public stakeholders. Since C3Grid makes use of Web Service technology, it is possible to switch from the current portal version to future version delivering a wider spectrum of possible portal entry points.

Two interfaces connect the portal to the rest of the C3Grid architecture. The first one is a Java API based web service interface to the Data Information Service (DIS). In order to achieve high performance in data search, the class libraries of the DIS are integrated directly into the portal framework. The user can freely define data queries by making specific selections for regions of interest, time constraints or variable names (based on the Climate Forecast Convention thesaurus) for all accessible distributed data sets in C3Grid and formulate data selection and pre- or postprocessing steps in a suitable workflow oriented way.

Furthermore the portal is connected as a client through a conventional axis web service to the Workflow Scheduling Service (WSS). Via this interface the user directly can specify data downloads and corresponding preprocessing functionality, as well as initiate the staging of data for later analysis in the workspace system described below. Aside from this rather elementary submission of scheduled jobs, the user can also submit complete predefined workflows (description see Sect. 6.3) or upload his dedicated workflow. He can start a workflow by choosing datasets and parameters for the analysis and eventually submit the job.

In the personal space of the portal, a list of all initiated downloads and submitted workflows allows monitoring and cancellation of jobs running. After a job is finished, the user gets an e-mail with a link to output data and graphics

---

[3] http://www.liferay.com

respectively as well as information, where the resulting output data is stored to be downloaded.

6.2.2 Data Information Service and Metadata

A precondition for uniform access is the definition of a common and consistent metadata system. In Sect. 4.4 an overview on existing standards is given. Until the start of the C3 project, the data providers had different metadata schemes for their data adapted to their special data profile, which could not be used directly for a common search over all available data. Based on the international schema required for describing geographic information and services (ISO 19115/19139) a specific C3 metadata profile was established.

It provides a general description framework for geographic data products in Earth System research. Some C3Grid adaptations and usage agreements were necessary to implement the standard for the C3Grid Community, but these are complementary and do not limit the international compatibility. They consist of C3Grid Data Access Interface References, Data Aggregation and Derivation Info and Data Content Description based on the Climate Forecast (CF) Convention. Further details see in Kindermann et al. (2007).

Data providers have to map their individual metadata schemes to that profile. Initial automated transformations of proprietary metadata formats into the common C3Grid metadata format were implemented at DKRZ (for data in the Climate and Environmental Retrieval and Archive (CERA) database system as well as the DKRZ file archive). Some data providers mapped their metadata manually or semi-automatically to the global C3Grid metadata profile.

The C3Grid data information service (DIS) uses the information from the metadata to discover the data provided in the grid. The DIS is based on the Open Archives Initiative Protocols (OAI-PMH) and Apache Lucene, which provides a fast full text search engine, displayed in Fig. 6.3. The algorithms for the required range search capabilities were optimized to enhance the performance of range queries. Being fully Java-based the DIS is directly integrated in the user portal to avoid performance losses. Details of the metadata harvesting framework PanFMP integrated in C3 can be found in Chap. 3.

Via a web service the portal user can issue Google-like queries for specific data sets, or query the index with physical variable names or select other constraints to the query he is interested in (see Schindler et al. 2007).

6.2.3 Collaborative Workspace and Data Management Service

C3Grid provides a secure and efficient environment for collaborative working on shared data. A Generation N Data Management System (GNDMS) was developed

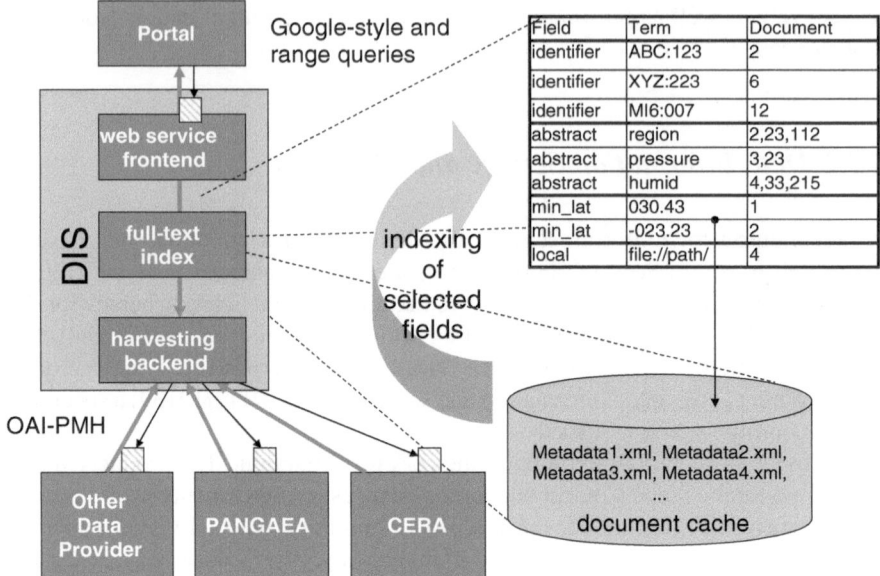

Field	Term	Document
identifier	ABC:123	2
identifier	XYZ:223	6
identifier	MI6:007	12
abstract	region	2,23,112
abstract	pressure	3,23
abstract	humid	4,33,215
min_lat	030.43	1
min_lat	-023.23	2
local	file://path/	4

Fig. 6.3 Architecture of the data information service (DIS)

at the Zuse-Institut Berlin (ZIB) in order to address data management aspects of community grids that need to provide distributed, computationally intense workflows with a data staging, replication and clean-up facility. GNDMS prepares, copies, replicates, caches, and deletes large datasets between supercomputer centers in an orchestrated and secure way. It abstracts from data sources via a data integration layer and provides logical names, data transfers via GridFTP, proper handling of GSI certificate delegation and workspace management. Whereas the first implementations were based on GT4, now only widely-used WEB protocol and interface standards such as REST and HTTP together with gridFTP are used, which makes the GNDMS very flexible. It can be deployed in any Servlet 2.5 compliant Application Container.

The task of the C3Grid data management is to provide means for making data available as input for grid applications. Actually, this comprises a number of different assignments: staging of data from primary data archives to local grid workspaces, transferring data between workspaces, delivering output data to the caller of the grid application and publishing results in the central data information service.

GNDMS addresses the special needs of climate computing and provides a number of unique features:

- metadata driven processing of climate data along the complete workflow chain
- support for data integration of nearly arbitrary data archives based on open protocols and ISO metadata
- an interface for republishing of processed data for intermediary reuse

- a highly customizable and extensible software package that has been integrated with established middleware standards
- storage management with automatic reclaiming of storage space
- a set of administrative tools for the configuration of single data management nodes and
- a directory service for the dynamic discovery of available data providers and storage locations.

GNDMS was developed for the C3Grid and is now being used in other grids too.

The layered architecture of DMS provides C3Grid with a failure resilient, secure, interoperable, manageable and extensible data management architecture that supports the unique compute-data-co-scheduling scheme of C3Grid. GNDMS allows unified data access to all participating data archives.

The C3Grid data management differs in some respect from common mainstream distributed data management systems. It actively supports its main client, the workflow scheduler, by performing its own planning of future transfers. For this task, it takes all information available about its current environment into account and corresponds with the scheduler to find agreements on transfer proposals.

The decision for exclusive access and management of grid workspaces as well as the design of a uniform interface to data archives are the key building blocks for making this possible. The latter interface keeps the type of storage system hidden from the grid environment but, nonetheless, provides a defined set of preprocessing operations which can be mapped, e.g., on queries of a relational database systems.

A system of naming conventions for consistent logical path names in the distributed C3Grid workspace was established in such a way, that the logical path name of a dataset is valid throughout the whole C3Grid Domain (i.e. Logical file abstraction, location independent) and mapped to its physical location only when necessary, e.g. when a processing task or transfer request needs to know the physical location for direct access to this data item.

The newest version of GNDMS allows access to data stored at a Earth System Grid data node. Thus, C3Grid becomes interoperable to the data federation used in the CMIP5 context. Scientists can now transfer data from ESG environment into the collaborative workspace of C3Grid and use them as input for further processing and analyzing steps. GNDMS uses CoG bindings and achieves interoperability with GSI, MyProxy and GridFTP from both Globus Toolkit 4 and the new plain Globus Toolkit 5.

6.2.4 Workflow Scheduling Service

For climate workflow processing, management, and scheduling, the C3Grid architecture comprises a central Workflow Scheduling Service (WSS) component. In C3Grid single tasks like data extraction, transfer, as well as data processing and analysis (model simulation) are considered as atomic entities, which can be interconnected in order to realize complex workflows. The WSS offers means to process such workflows automatically. It simultaneously tries to optimize response time and data management aspects during execution by dynamically selecting adequate resources and transfer routes within C3Grid. Internally, each submitted workflow is represented by a management service instance, which handles all global execution aspects like task interdependencies, execution site selection, transfer planning, and situation dependent scheduling. Single tasks and their specific execution are then managed by task execution services.

The elementary components of a workflow are called atomic tasks. They are described via Job Submission Description Language (JSDL) containing staging, execution and publishing directives. In order to describe workflows consisting of several sequential but mutual dependent chains of elementary tasks, a C3Grid-specific Workflow Specification Language (WSL) was defined which describes dependencies between atomic tasks.

Climate research requires the execution of complex modular applications. The scheduling system must be able to handle sets of interdependent tasks which comprise multiple steps in the analysis of a scientific problem. Moreover, high volume data requirements emerge from long-term experimental recordings or simulation runs. Therefore, the scheduler has to work in cooperation with the data management system which has the full information about the location of data and their replicas. Sophisticated strategies for the orchestration of data availability and compute resource usage have to be applied in order to allow seamless co-allocation of resources for user workflows. The refined interaction of the WSS with the GNDMS is based on an abstract negotiation protocol (see Grimme et al. 2007).

6.2.5 Provider Layer: Data Archive and Compute Resources

Data providers publish local data archives for use in C3Grid. Setting up a new provider requires installing and configuring the C3Grid data management software GNDMS and publishing descriptive ISO 19115 metadata for harvesting by the data information system. The data access strategy is based on standard web service technology and a community interface which specifies the users' time, space and variable constraints for the requested data as well as additional processing functionalities supplied by the provider. The provider specific implementation of the data access as well as of the data preparation is hidden behind a common interface. Following this general C3Grid Data Access Interface, a new data provider has to implement the specific actions for data access and preprocessing functionality. Access to provider data archives may be setup to either use a simple shell-based API or provider-specific GNDMS plug-in.

As C3Grid offers also postprocessing by diagnostic workflows compute resources are included in the C3Grid infrastructure. They receive file transfer requests from the C3Grid data management system as workflow input data, batch requests from the C3Grid scheduler for workflow execution or file transfer requests for output data delivery to the C3Grid portal. Therefore compute providers have to install and maintain the necessary infrastructure for these services.

6.3 Workflows

Data analysis in a reproducible way builds up a workflow of well defined process steps. For a certain problem class the succession of specific process steps is typical and will be repeated by the scientists with many various data sets. A main feature of C3Grid is to provide a possibility to make easy use of generalized forms of such analysis workflows. Depending on the load at the various compute sites in C3Grid the job is scheduled to run on a suitable resource. The negotiation process to choose the respective compute resource depends not only on the load of the resource, but takes also into account, how the data needed for the workflow can be transferred to the compute provider up to the planned starting time of execution. As a proof of concept, some typical workflows for analysis are implemented, which serve as templates for future extensions. The portal offers the implemented workflows by clicking the respective tab and the user can then input the relevant parameters to control the job run.

Up to now, as examples the following workflows were implemented:

- Storm track analysis: This tool is widely applied in order to estimate the intensity and spatial distribution of baroclinic wave activity in the mid-latitudes. It is thus, for example, helping to identify differences between different forcing scenarios or different models using an indicator quantity for complex atmospheric processes. It can be applied to different vertical levels available in an atmospheric dataset, thus providing a basis for an investigation of the 3-dimensional structures of signals or differences.
- QFlux: Humidity fluxes are a key parameter for the understanding of the variability associated with precipitation, including the occurrence of extreme events like extremely high rainfall amounts in different time scales, or drought. As fluxes must be considered in different vertical levels and height ranges for the diagnosis, the tool provides possibilities for a vertical integration, including an interpolation using surface layer values.
- CAPE: Convective Available Potential Energy is a measure describing the potential of an observed or simulated atmosphere to generate strong convection, which is related to the occurrence of local rainstorms, hail or downdrafts which can produce large damage.
- Visualization: A basic visualization module is based on Grid Analysis and Display System GrADS[4] and produces 1D and 2D plots as well as animations. In it's simpliest form the workflow allows to visualize original or preprocessed NetCDF data from data providers. Due to its generic character, the module can also be

---

[4] http://www.iges.org/grads/

used as the final step in analysis workflows. Furthermore, a concept of High End Visualization on distributed resources with user interactions was developed. The implementation of these improved visualization capabilities is still under work.

- Basic ensemble statistics: Typically, simulations are carried out in several different realisations to lower the uncertainty of model runs. The workflow computes statistical values over an ensemble of input files. Depending on the actual operator the statistical measure (minimum, maximum, sum, average, variance, standard deviation, or a certain percentile) over all input files is calculated. The module uses the Climate Data Operators cdo,[5] a tool set for analysis and processing of climate and Numerical Weather Prediction model data.

- Global Information System (GIS): This workflow provides an additional output format for geographic information systems. It is designed to serve as the final step in analysis workflows to convert the calculated output files into a format which can afterwards be read in by a local GIS program (e.g. ArcGIS or GRASS) smoothly. For transformation of NetCDF into ASCII the Geospatial Data Abstraction Library GDAL is used.

- Circulation Weather Types (CWT): This diagnostic workflow calculates ten mean-sea level pressure patterns that represent the main circulation weather types (CWTs) for a target region. The version of CWTs in the present diagnostic workflow is an aggregated set of the so-called Lamb CWTs (Jones et al. 1993). As input the user specifies a center point and sixteen surrounding grid points, where the horizontal geostrophic wind components, as well as the geostrophic relative vorticity is derived from the mean-sea level pressure field. Then a classification of weather types can be processed for every time step.

- Cyclone Tracking: Like the workflow before, this diagnostic module analyses the mean sea level pressure field for identifying and tracking the occurrence of surface low pressure systems (i.e. cyclones). Positions of cyclone centers for each time step are calculated and tracks from the identified cyclones are compiled (Murray and Simmonds 1991). Based on these cyclone tracks climatological cyclone statistics can be inferred.

- Grass reference evapotranspiration: Results of climate research concerns not only the climate community itself, but also a broader user group which is not interested in the original raw data, but in climate indices and derived parameters. The grass reference evapotranspiration was implemented in C3Grid as a representative of this class of new workflows for users outside the climate research community. It calculates the grass reference evapotranspiration in a standardized way (see Wendling et al. 1991).

These standard workflows for easy analysis of data have served as a proof of concept for the gridification of complex diagnostic tasks and will be supplemented by further workflows in the near future. For experienced users C3Grid provides an interface for the upload of private workflows which can be submitted in the grid infrastructure to make use of the distributed environment.

---

[5] https://code.zmaw.de/projects/cdo

6.4 Status and Perspectives

The C3Grid infrastructure offers a collaborative platform for scientists in climate research. When a user is browsing in the central metadata catalog, the overview descriptions of all available datasets from all participating data provider sites are presented. After submitting the data request of the special data selected the according data and metadata are transferred to the local grid workspace. If necessary some preprocessing operations (like cutting out a special region or extraction of a set of variables) are performed and the resulting data is made available for the user (see Fig. 6.4).

A common description by a new defined metadata profile based on international ISO standards allows discovery of very heterogeneous data in distributed archives. A common interface between grid and data provider hides the different modes of data access and eases this process for the user. C3Grid spans an umbrella over data provider with various offerings of different data types. Beside model results (e.g. at WDC-Climate and University of Cologne/Freie Universität Berlin) the user finds also observational data, campaign data from WDC-MARE, and data products resulting from processed satellite data (WDC-RSAT). By means of providing

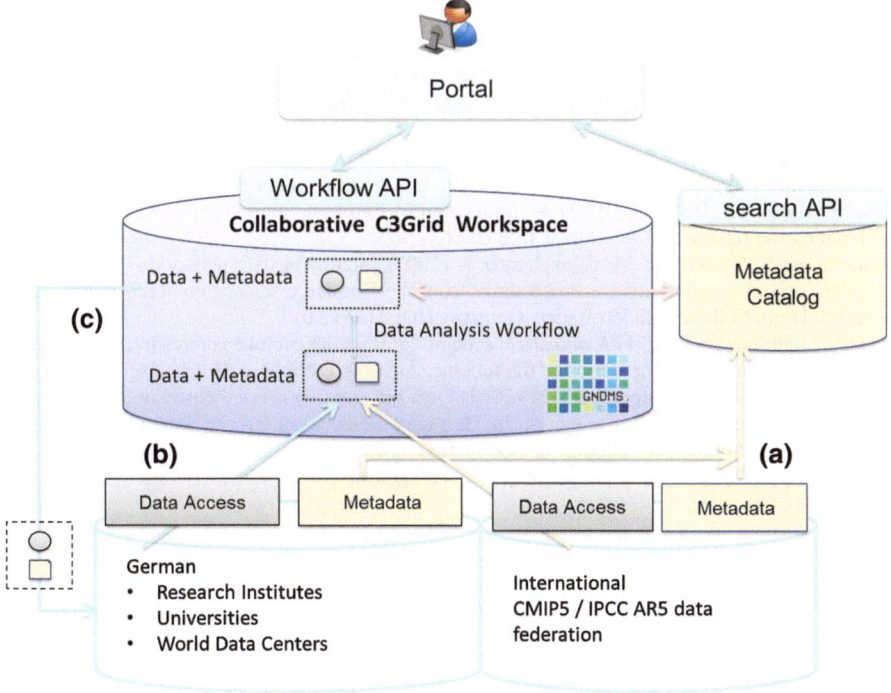

Fig. 6.4 Data access from C3Grid data provider and ESG data nodes into the collaborative workspace

enhanced preprocessing functionality locally at data provider sites the C3Grid solves the problem that the extracted data volumes from the archives are often of such a size that they cannot be easily downloaded due to limitations in data transfer capabilities. C3Grid has thus established a very useful alternative to the "download and process at home" paradigm currently in place because the preprocessing functionality minimizes the amount of data to be actually transferred. Furthermore the provision of scheduled compute resources allows to run diagnostic workflows on several sites. Typical workflows for data analysis were defined and implemented as proof of concept. The portfolio of diagnostic tools is increasing and will include more analysis tools in the future.

Additional data providers will be integrated in the C3Grid to broaden the data basis for the scientists. Similarly, further compute resource providers will join the grid. First experiences using virtual machines as compute resources were gained in the previous months and show promising results. By means of virtualization templates for the provision of compute resources the process of deploying new compute resources properly instrumented with software is greatly simplified.

With the enhancement of GNDMS connected directly to ESG data nodes, the data pool becomes significantly larger. For C3Grid this is an important step in the development from a national German grid infrastructure to an internationally linked collaborative research environment for climate science.

References

Grimme C, Langhammer T, Papaspyrou A, Schintke F (2007) Negotiation-based choreography of data-intensive applications in the C3Grid project. In: German e-science conference, Baden-Baden

Jones P, Hulme M, Briffa K (1993) A comparison of Lamb circulation types with an objective classification scheme. Int J Climatol 13:655–663

Kindermann S, Stockhause M, Ronneberger K (2007) Intelligent data networking for the earth system science community. In: Proceedings from the German e-science conference 2007, Max Planck Digital Library, Baden-Baden, Germany (ID: 316512.0)

Murray R, Simmonds I (1991) A numerical scheme for tracking cyclone centres from digital data. Part I: development and operation of the scheme. Aust Meteorol Mag 39:155–166

Schindler U, Bräuer B, Diepenbroek M (2007) Data information service based on open archives initiative pro-tocols and apache lucene. In: German e-Science conference, Baden-Baden

Wendling U, Schellin HG, Thomä M (1991) Bereitstellung von täglichen Informationen zum Wasserhaushalt des Bodens für die Zwecke der agrarmeteorologischen Beratung. Z f Meteorologie 41:468–475

Chapter 7
Earth System Grid Federation: Federated and Integrated Climate Data from Multiple Sources

Dean N. Williams , Gavin Bell , Luca Cinquini , Peter Fox, John Harney and Robin Goldstone

7.1 Background

In climate science, the quantity of data in use by 2020 is expected to be in the hundreds of exabytes[1,2] (EB, where 1 exabyte $= 1018$ bytes). Current and future heterogeneous climate data will be distributed around the globe and must be harnessed to find solutions to mission-critical problems. Additionally, more requirements and more con-

---

[1] CCDC Workshop, International Workshop on Climate Change Data Challenges, June 2011, http://www.wikiprogress.org/index.php/Event:International_Workshop_on_Climate_Change_Data_Challenges.

[2] CKD Workshop, Climate Knowledge Discovery Workshop, March 2011, DKRZ, Hamburg, Germany, https://redmine.dkrz.de/collaboration/projects/ckd-workshop/wiki/CKD_2011_Hamburg

---

D. N. Williams (✉) · G. Bell
Program for Climate Model Diagnosis and Intercomparison, Lawrence Livermore National Laboratory, Livermore, USA
e-mail: williams13@llnl.gov

G. Bell
e-mail: bell51@llnl.gov

L. Cinquini
NASA Jet Propulsion Laboratory, Pasadena, USA
e-mail: luca.cinquini@jpl.nasa.gov

P. Fox
Rensselaer Polytechnic Institute, Troy, USA
e-mail: pfox@cs.rpi.edu

J. Harney
Oak Ridge National Laboratory, Oak Ridge, USA
e-mail: harneyjf@ornl.gov

R. Goldstone
Lawrence Livermore National Laboratory, Livermore, USA
e-mail: goldstone1@llnl.gov

W. Hiller et al., *Earth System Modelling – Volume 6*, SpringerBriefs in Earth System Sciences, DOI: 10.1007/978-3-642-37244-5_7, © The Author(s) 2013

straints are needed to expand and integrate new modeling capabilities and tasks, such as climate prediction, uncertainty quantification (UQ) of model performance, test-bed development, and assimilation of more diverse data sets. These data exploration tasks can be complex and time-consuming, and frequently involve numerous resources spread throughout the modeling and observational climate communities. Staff expertise and core competencies must therefore be flexibly applied to multiple projects and programs to accommodate more complex applications and state-of-the-art science analysis, while allowing resources to be adapted to address future areas of interest in climate research. To process state-of-the-science models and analyze those results, researchers will need more complex and flexible architectures that can run heterogeneous applications over fast heterogeneous networks. Performing remote operations will reduce data movement and minimize the amount of data to be stored. By working closely with the community, the U.S. Department of Energy (DOE) Office of Biological and Environmental Research (BER) is exploring and developing hardware and software workflow applications to integrate DOE's climate modeling and measurements archives. BER is coordinating efforts to develop infrastructure for national and international model and data comparisons. Having an integrated infrastructure (or framework) in place will allow climate science centers worldwide to deploy a wide range of data visualization, diagnostics, and analysis tools with familiar interfaces—a critical issue for building data systems that process very large, high-resolution climate data sets and meet the growing demands of this intensely data-rich community.

Because the infrastructure for this type of community-wide network must be interoperable, each system must be established on a standard set of services, application programming interfaces (APIs), and protocols so that other systems can interconnect their components. By encapsulating component service operations behind message-oriented service interfaces, users will be isolated from the details of implementations and distributed service locations and freed to work in a virtual workspace, if so desired.

To this end, DOE has made elegant architectural investments in the Grid Forum: (1) Grid Computing and (2) Data Grids. In the area of Grid Computing, the Open Science Grid (OSG) successfully engages a variety of science domains in adapting their software components to use a distributed set of DOE and community-wide computing and storage resources. In the area of Data Grids, the Earth System Grid Federation (ESGF) (shown in Fig. 7.1) successfully provides distributed data systems and services to the climate-modeling community.

ESGF is an exemplary use case showing how to achieve an infrastructure with open, common-component architecture for distributed science data systems and services. For this infrastructure to have continual success, we must identify the overlapping needed services that are common across project domains and design the system so that the integration works well and components are reusable, scalable, and extensible. As technology changes, the infrastructure must be flexible enough to evolve and keep pace with the growing demands of climate science.

Managing high volumes of climate data within a network of high-performance computing (HPC) environments presents unique challenges. How data are orga-

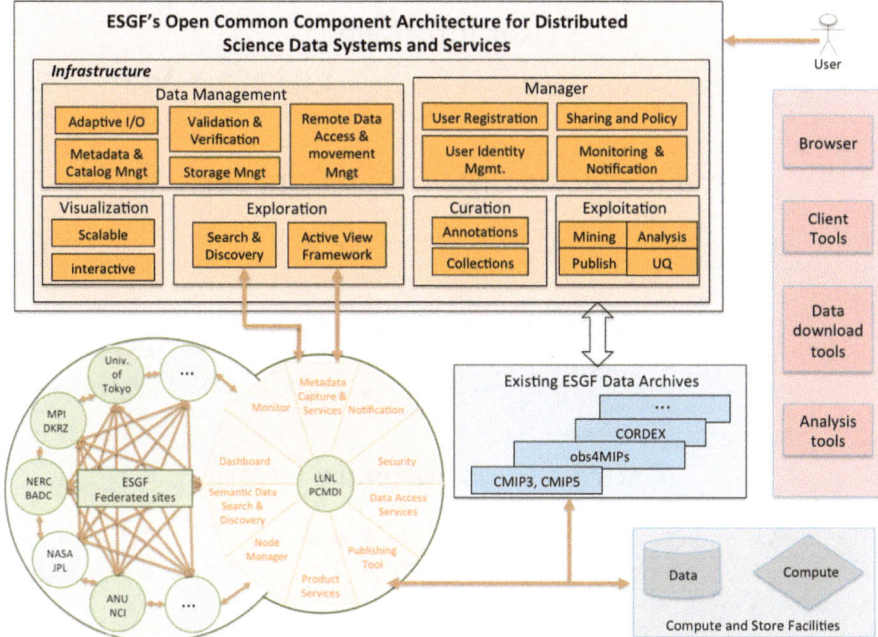

Fig. 7.1 ESGF's federated framework integrates distributed data systems and services for discovery-class research that explores cross-cutting climate science domains. The *dark orange boxes* are common component services needed for the distributed data systems. Communications between the components are implemented via a standard set of APIs and protocols defined by the science community. ESGF currently comprises over two-dozen nodes, and five of these (*indicated in the lower left by a darker shade of green*) host replicas of a substantial number of the CMIP5 data sets (i.e., PCMDI, DKRZ, BADC, NCI, and the University of Tokyo). Users have access to all data throughout the federation, regardless of which ESGF node is used

nized can have considerable impact on performance. Often, the only realistic choice for storage device is robotic tape drives within a hierarchical management system. In the worst case, poor data organization means that some data may never be accessed simply because it takes too long (i.e., storage access, compute resources, network). Users generally store the data themselves, possibly without being aware of how to most effectively use the hierarchical storage system. Additionally, some applications require that large volumes of data be staged across low-bandwidth networks simply to access relatively small amounts of data. Finally, when data usage changes or storage devices are upgraded, large data sets may need to be reorganized and quite possibly moved to a new location to take advantage of the new configuration. To address these concerns and others, the climate data community needs a network architecture that offers more intelligent and complete layered data services (as shown in Fig. 7.2), providing users with increased information about the data, its anticipated usage, storage requirements, and network system characteristics. Such a system will include a layered service structure that is invisible to users but that effectively manages the system to ensure a truly efficient, productive workflow.

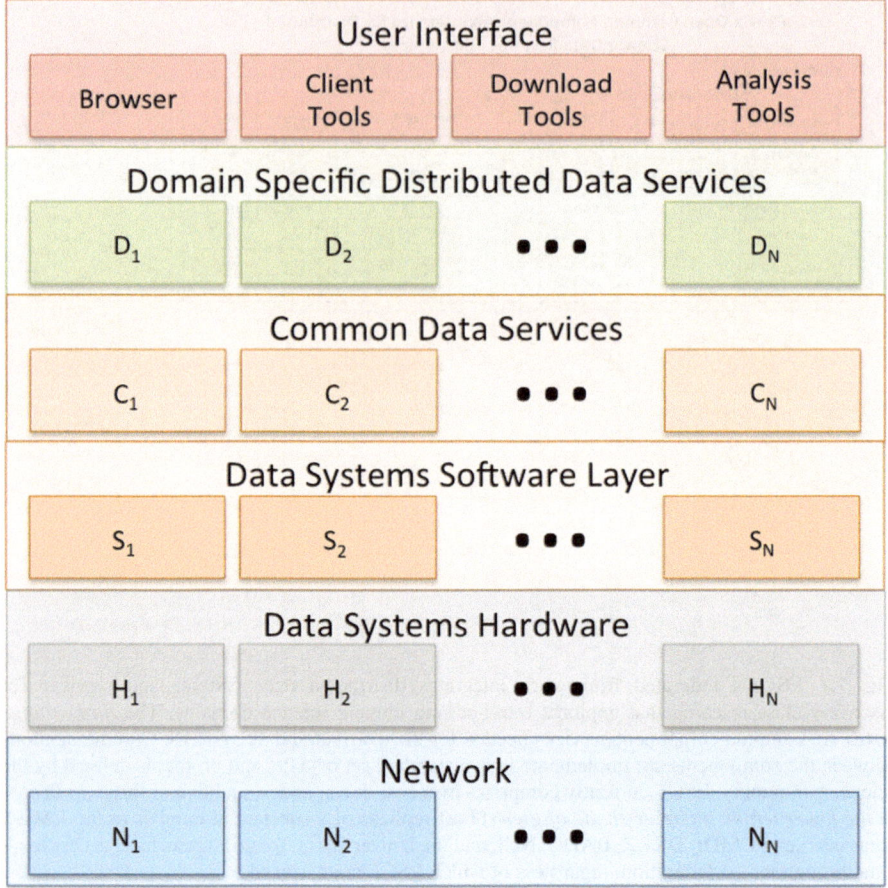

Fig. 7.2 The diagram depicts the service layers hidden to the user. Standard APIs and protocols define the communication between each layer

- **Domain-Specific Distributed Data Services**: At this level of the hierarchy, application components required for specific climate projects are clearly defined. For example, if a climate project has unique node protocol services for managing its distributed data system worldwide, those common components will fall into the first box of Fig. 7.2. These services are specific to the domain space for this particular climate data project. D1 to DN captures the set of specific services for each project domain.
- **Common Data Services**: These are services that all project domain areas can use, such as movement, curation, discovery, annotation, and exploration. The C1 to CN layer exhibits standard protocols or standard interfaces from one layer to the next, allowing for extensibility and reuse.

- **Data Systems Software Layer**: The lower layers of the hidden services are closer to hardware and thus require definitions for more specific services. S1 to SN concerns itself with metadata, file size, provenance, and workflow.
- **Data Systems Hardware**: The H1 to HN layer represents hardware, such as clusters, clouds, and in situ data analysis for large-scale computational data analysis and modeling. At this level, interfaces must be defined to communicate with machines throughout the evolution of a simulation, for example, during the complete calculation for an uncertainty quantification analysis.
- **Networks**: Binding the collection of disparate hardware components, resources, and users are the networks. The N1 to NN represents high-speed (or low-speed) networks required to replicate large data holdings at storage facilities and to federate connectivity.

Given the critical importance of scientific climate data and its projected size by 2020, the climate research community must continue to specify a format for common activities as well as standards, APIs, and protocols to facilitate the development of infrastructures (such as ESGF) that support the community's long-range efforts. We cannot afford to work in an ad hoc fashion without proper standards for building hardware and networks or bonding software together via the specific protocols because doing so will cost the DOE BER and the climate community at large considerable time and resources. If DOE and funding agencies around the world are to optimize their investment in data, they must ensure that a common open architecture is in place and a significant fraction of that architecture is shared among the different climate activities, rather than having specific domain architecture for each project.

7.2 Science Drivers

The Earth System Grid (ESG) was established in 1999, to meet the needs of modern-day climate data centers and climate researchers. Specifically, ESG addresses the requirements of data centers and climate researchers for interoperable discovery, distribution, and analysis of large and complex data sets. Under the leadership of the DOE BER Program for Climate Model Diagnosis and Intercomparison (PCMDI) at Lawrence Livermore National Laboratory (LLNL) and in partnership with Argonne National Laboratory (ANL), the National Center for Atmospheric Research (NCAR), Lawrence Berkeley National Laboratory (LBNL), Oak Ridge National Laboratory (ORNL), Los Alamos National Laboratory (LANL), the National Aeronautics and Space Administration (NASA), the National Oceanic and Atmospheric Administration (NOAA), and others in the national and international communities—including centers in the United Kingdom, Germany, France, Italy, Japan, and Australia—an internationally federated, distributed data archival and retrieval system was established under the name Earth System Grid Federation, or ESGF. Although this development effort is coordinated internationally, the ESG team is the primary contributor

Table 7.1 ESGF distributed data archive

Type	Federated Data Sets (i.e., Projects)
Model	Phases 3 and 5 of the Coupled Model Intercomparison Project (CMIP3 and CMIP5)
Model	Coordinated Regional Climate Downscaling Experiment (CORDEX)
Model/Observational	Climate Science for a Sustainable Energy Future (CSSEF)
Model	European Union Cloud Intercomparison, Process Study & Evaluation Project (EUCLIPSE)
Model	Geo-engineering Model Intercomparison Project (GeoMIP)
Model	Land-Use and Climate, Identification of Robust Impacts (LUCID)
Model	Paleoclimate Modeling Intercomparison Project (PMIP)
Model	Transpose-Atmospheric Model Intercomparison Project (TAMIP)
Observational	Clouds and Cryosphere (cloud-cryo)
Observational	Observational Products More Accessible for Coupled Model Intercomparison (obs4MIPs)
Model	Reanalysis for the Coupled Model Intercomparison (ANA4MIPs)
Model	Dynamical Core Model Intercomparison Project (DCMIP)
Model	Community Climate System Model (CCSM) / Community Earth System Model (CESM)
Model	Parallel Ocean Program (POP)
Model	North American Regional Climate Change Assessment Program (NARCCAP)
Model	Carbon Land Model Intercomparison Project (C-LAMP)
Observational	Atmospheric Infrared Sounder (AIRS)
Observational	Microwave Limb Sounder (MLS)
Observational	Modern-Era Retrospective Analysis for Research and Applications (MERRA)
Observational	Tropical Rainfall Measuring Mission (TRMM)
Observational	Multi-angle Imaging SpectroRadiometer (MISR)

to the ESGF software stack. ESGF work has resulted in production of an ultra-scale data system, empowering scientists to engage in new and exciting data exchanges that could ultimately lead to breakthrough climate-science discoveries.

ESG was critical to the successful archiving, delivery, and analysis of the Coupled Model Intercomparison Project (CMIP), phase 3 (CMIP3), which provided data for the Fourth Assessment Report (AR4) of the Intergovernmental Panel on Climate Change (IPCC). It is proving to be equally important in meeting the data management needs of CMIP, phase 5 (CMIP5), which is providing petascale data informing the 2013 IPCC's Fifth Assessment Report (AR5). Although ESGF has been indisputably important to CMIP, its current and future impact on climate is not limited only to this high-profile climate project. ESGF has been used to host data for a number of other climate projects, including CCSM, CESM, NARCCAP, C-LAMP, POP, and AMIP. These data archives have been augmented with observational data sets (for example, ARMBE, CDIAC, NASA satellite observation data sets (CloudSat, MLS, MISR, AIRS, and TRMM), and NASA-NOAA reanalysis data sets (MERRA, CERES)).

Because of rapid increases in technology, storage capacity, and networks and the need to share information, communities are providing access to federated open-source collaborative systems that everyone (including scientists, students, and policymakers) can use to explore, study, and manipulate large-scale data. The ESGF software stands out from these emerging collaborative knowledge systems in the climate community along multiple dimensions: the amount of data provided (petabytes), the number of global participating sites (dozens), the number of users (over 25,000), the amount of data delivered to users (over 2 petabytes), and the sophistication of its software capabilities. ESGF is therefore considered the leader for both present and future data holdings as shown in Table 7.1.

7.2.1 Typical ESGF Facility: PCMDI LLNL

Figure 7.3 shows all PCMDI/ESGF components. The low-bandwidth front-end servers reside on the LLNL enterprise network while high bandwidth data access is provided via Data Transfer Nodes (DTNs) located in the Science DMZ.[3] Multiple petabytes of climate data are stored in the Climate Storage System (CSS), a high performance SAN storage environment. The CSS has private network connections to all other components: the front-end servers, the DTNs, and the back-end simulation/analytics cluster.

In order to meet high-bandwidth needs for moving large volumes of climate data for international collaboration, LLNL is currently the only DOE National Nuclear Security Administration laboratory with a 100-giga-bit-per-second (Gbps) routed endpoint on the new ESnet5 100 Gbps backbone. Multiple DTNs with 10 Gbps interfaces will utilize the parallel capabilities of GridFTP to achieve maximum throughput across this 100 Gbps connection.

7.2.2 Enhancing the Process of Science

As the demand for robust and consistent scientific data distribution platforms increase, user interface (UI) design and implementation have become one of the more important tasks in the creation of a scientific data access portal. Potential data consumers, specifically end users of the ESGF portal, are inevitably concerned with the manner in which services (such as node management, search, and login) are presented. For the most part, end users discover and access data via the ESGF front-end UI application. The ESGF node team has incorporated a number of components to the front-end design to accommodate a wider community of potential users that may be interested in various data sets (e.g., scientists interested in comparing modeled output with observational data). As with most modern search portals, text search has

---

[3] http://fasterdata.es.net/science-dmz/science-dmz-architecture/.

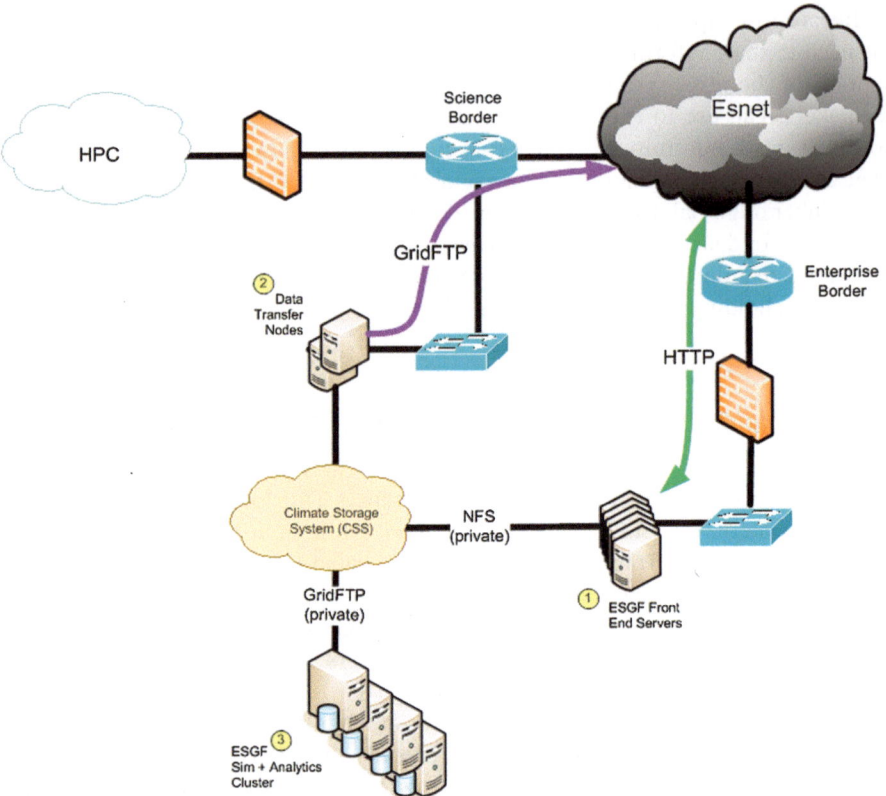

Fig. 7.3 ESGF data flows: (*1*) Users communicate with ESGF front-end servers on the LLNL enterprise network via HTTP. Small data sets are retrieved by the front-end servers from the CSS and returned to the user via HTTP. (*2*) Large data sets are made available to users via GridFTP from the CSS Data Transfer Nodes (DTNs) located in the LLNL Science DMZ. Firewall bypass by DTNs ensures good/consistent performance for these large file transfers. (*3*) ESGF may perform analysis of raw data if requested by users through the front-end servers. Analysis jobs are dispatched to the ESGF Sim+Analytics cluster, which retrieves necessary data from the CSS, performs analysis and puts results back in the CSS, where they can be retrieved by the customer via GridFTP

been enhanced with autocomplete capabilities to aid users who may be unsure of specific search strings. The temporal search tool leverages the highly effective range search capabilities of the Apache Solr search back end, allowing users to extract data sets according to ranges constrained by time measurements via Sol's Lucerne-based inverted-index technology. The faceted-based category navigation tools were expanded to include flexible terminology (for example, "instruments" in observational terminology) while providing direct support for the structured terminology of the climate community.

As shown in Fig. 7.4, UI components have been developed for the key functional areas listed below. Significant refinement has been achieved in each component area based on production use of the system and user feedback. The design goals for

Fig. 7.4 The ESGF node installation script deploys the home page of a minimally configured web front-end application, as shown to the *left*. Included on the home page are search capabilities, information pertaining to the node and organization, quick links to commonly used tools, links to other ESGF nodes, and analysis tools. Example search page and analysis/visualization results are shown on the *right*

UI tools include exposing the base system functionality, providing a consistent and intuitive user experience, and supporting a flexible and maintainable framework for future enhancements and revisions.

- **The home page** provides visitors with general information about the discipline-specific portal, starting points for discovering data collections, direct access to notable data collections, important notices regarding system status, and access to login and account request functions. The home page allows a climate project to customize its format to include project-specific information, data browser entry points, logo images, and color palette.
- **User registration** offers a multi-step workflow for account creation, approval, and validation. The resulting account may be used to authenticate any user at any portal in the federation.

- **User and group management** allows registered users to change account settings and request access to privileged data collections. It also provides tools for group administrators to approve group requests and manage group membership.
- **Login** allows registered users to authenticate with the federated system with an OpenID user identifier. Users may request password delivery via e-mail in case of lost credentials.
- **Data browsing** provides support for file system-like hierarchies and high-level associative arrangements such as experiment, and project-related listings.
- **Data search** is the primary data-discovery method for most users. This component provides a simple and familiar text-based search as well as faceted navigation for exploration-based metadata inquiry.
- **Data download** allows for individual file the download via hyperlink and for bulk download requests from the file-listing interface using generated WGET scripts and the Globus Online-hosted data movement service. If data collections are restricted and under access control, the user is directed to authenticate prior to data download.
- **Data transfer** allows registered and authorized users to request and manage groups of files from deep storage systems throughout the federation. Users can access real-time status reports and are notified by RSS feeds or via e-mail when transfers are complete.
- **Data visualization and subsetting** provides an interface for requesting charts, plots, and data subset downloads. Users may choose variables of interest and select sub-regions geospatially using an interactive map and temporally using time controls.

Users can also interact with the ESGF distributed archives for knowledge discovery via analysis tools. Funded under BER in support of its science mission, the Ultra-scale Visualization Climate Data Analysis Tools (UV-CDAT) framework is designed to integrate six analytical and visualization tools—CDAT, VisTrails, ParaView, VisIt, DV3D, and R—all under one application. Based on Python, it links disparate software subsystems and packages to form an integrated environment for analysis. UV-CDAT's design and openness permit the shared development of climate-related software by the collaborative community. In particular, the goals of the UV-CDAT project are to (1) prepare for the CMIP5 data archive and assessment process by developing derived data products and user-reproducible workflows and analysis archives; (2) develop capabilities to inter-compare ungridded observational data sets and model data for validation; (3) deliver efficient scalable analysis and visualization for high-resolution simulation data; (4) deliver data products in formats suitable for expert and non-expert users; and (5) build all capabilities on existing ESGF node infrastructures.

Figure 7.5 shows the new interactive UV-CDAT graphical user interface (GUI) in the background and the UV-CDAT ESGF node search and browse GUI. These features allow users to search and browse the ESGF distributed archive from within the UV-CDAT analysis tool as if they were on a web browser. Once a data set is located, a user can download it directly to the UV-CDAT application.

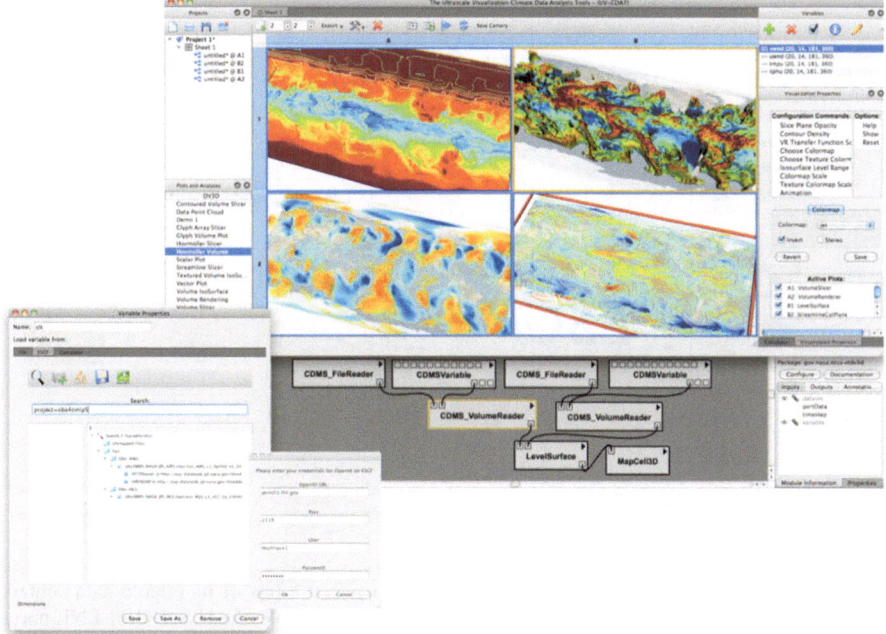

Fig. 7.5 UV-CDAT searches and accesses the ESGF node archive at LLNL/PCMDI (shown in the *lower left*) and displays the requested results using DV3D (in the *upper right*). The UV-CDAT reproducible workflow is displayed under the four-panel visualization spreadsheet

The interaction between users and the UV-CDAT GUI is also depicted in Fig. 7.5. A user interacts with the UV-CDAT GUI by invoking scripts, clicking buttons, or dragging variables and plot types. In response to these actions, UV-CDAT records a series of operations and converts them into provenance-enabled workflow operations that allow the user to share work with others as well as to reproduce the operations.

7.3 Data Quality and Publishing in an International Setting

ESGF seamlessly joins climate science data archives and users around the world. As shown in Fig. 7.6, it accesses many wide-area networks (WANs) to remotely connect researchers, policymakers, and other users to climate data projects through web-based interfaces and analysis tools (as described in Sect. 7.2.2). Data providers make data available to the federation by publishing to one of two-dozen ESGF node portals. Data can be replicated at other ESGF node sites for backup, to improve ease of use, or to exploit site resources. In the process of ESGF node replication, data are moved via GridFTP or via the HTTP protocols. This process is the same for user data movement.

Fig. 7.6 Federation of ESGF showing collaborations between a few of its remote data centers, data archives, and potential data transfers between sites. For example, the U.S. DOE/LLNL portal (at the *top*) harvests IPCC/CMIP5 data from 10 countries. That is, the original data resides at the data centers, but subsets of the data are replicated at LLNL for backup, better access and use. The DOE/LLNL portal URL is http://pcmdi9.llnl.gov

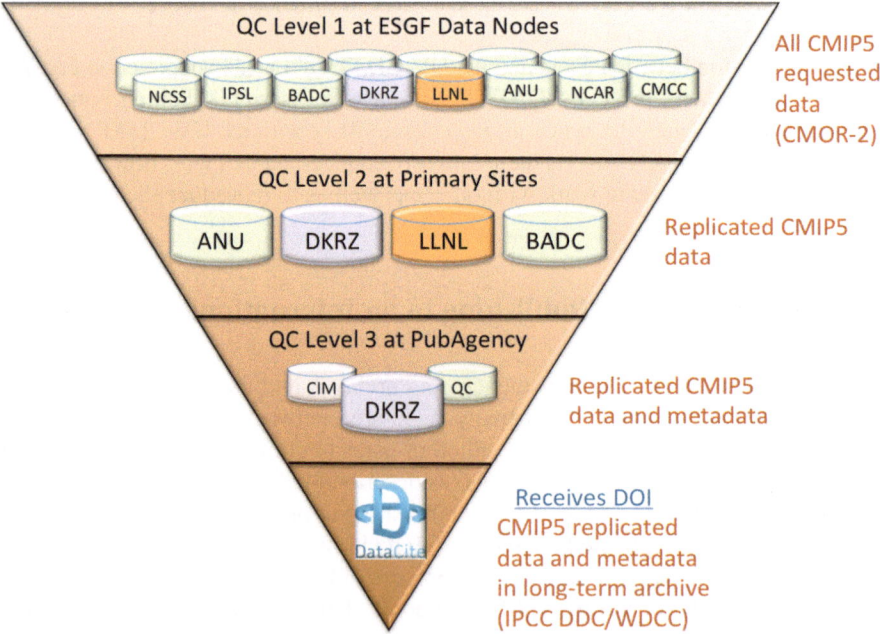

Fig. 7.7 Three-layer quality assurance concept

Part of the process of publishing and replicating data is the data quality control check operations, which ends in digital object identifiers (DOIs). This process takes a three-layer approach to data-quality assurance, as shown in Fig. 7.7. When a modeling or data center publishes data to ESGF, the system performs the automatic Quality Control (QC) check level 1 on the data. This QC check confirms that the data are in compliance with the NetCDF and Climate Forecast (CF) convention. For the second-level QC check, subsets of data are transferred to either DKRZ or LLNL, where a quality control code is run on the data to check for consistent application of units, measurements, etc. In some cases, visual inspection of the data also takes place for correctness. Once the data are accessible to the user community and have been used without complaints for a three-month period, they are elevated to QC level 3 status and issued a DOI. DOI data sets are then replicated to the IPCC Data Distribution Centre (DDC) at the World Data Center for Climate (WDCC) and to long-term archive at DKRZ and BADC. This process will occur over the next two years.

7.4 Data Intensive Climate Science

Rapid advances in experimental capabilities, networks, hardware, computational technologies, and techniques are driving exponential growth in the volume, acquisition rate, variety, and complexity of scientific data. This new wealth of meaningful data has tremendous potential for scientific discovery. However, if scientists are to use this vast resource to achieve scientific breakthroughs, the data holdings must be exploitable so that the information can be analyzed effectively and efficiently, and the results shared and communicated easily. The explosion in data complexity and scale makes these tasks exceedingly difficult to achieve—particularly given that an increasing number of climate projects are working across techniques, integrating simulations with experimental or observational results. Consequently, we must continually build on ESGF's data-management, analysis, and visualization tools to provide research teams with easy-to-use, end-to-end solutions. These solutions must facilitate (and where feasible, automate) every stage in the data lifecycle, from collection to management, annotation, sharing, discovery, analysis, and visualization. Hereby, a core set of ESGF functionalities must be offered to all climate projects, but individual processes will require customization so they can be adapted to a project's specific needs and fit into the different research and analysis workflows.

Therefore, we will leverage existing DOE and community-proven core technologies and facilities so that we can provide an even more comprehensive portfolio of data-management, analysis and visualization capabilities to the entire climate community. We will build on technologies developed within DOE-funded projects—such as the ESGF, UV-CDAT, Globus Online, CSSEF test bed, Ensemble Data Analysis Environment, and Extreme Scale Visual Analytics—and adapt and extend these tools in collaboration with the DOE application partnerships (U.S. and international agencies, universities, and private companies).

In addition to our trademark advances in data preparation, search and discovery, data access, security, and federation (as mentioned above), over the next few years, we will focus on expanding into the new areas of data mining, provenance and metadata, workflows, HPC, data movement, and data ontology. Using projections of upcoming scientific endeavors, we can extract and summarize the high-level requirements that we plan to address for ESGF, as shown in Fig. 7.8.

In the coming years, the process of conducting data-intensive climate science research will remain primarily the same as described in Sect. 7.2.2; however, data processing will be more commonly performed at remote data centers. The goal is to have UV-CDAT analysis processes co-located where the data reside. Through UV-CDAT, provenance metadata would be recorded at every step in the process and archived as a workflow configuration co-located with the data product. Later, other scientists could run the same analysis from the workflow descriptor to confirm the results, and they could expand on early findings by running different variations of a processing algorithm or using different input data sources.

Figure 7.9 illustrates one view of the planned infrastructure for ESGF. In this future setup, key data centers, identified as Leadership Computing Facilities, are located throughout the federation, providing users with ready access to the complete data archive. A ParaView cluster performs the calculations requested for different projects and offers users multiple display options for viewing the returned results. All data products and workflow descriptors in this planned ESFG infrastructure would be automatically archived to improve the ease of sharing knowledge, both about the climate predictions and the data-analysis applications.

7.5 Data, Workflow, Middleware Tools, and Services

ESGF delivers a comprehensive, end-to-end and top-to-bottom environment for current and emerging exascale climate science, as shown in Fig. 7.10. We emphasize data services at each level of the architecture. Figure 7.10 is an expanded view of Fig. 7.2, showing in greater details the hidden-layer services along with the analysis and visualization services accessible to users. More than a proof-of-concept, the production of ESGF for climate projects is evidence that a distributed dynamic federated system is flexible enough to support a wide range of climate projects by providing the following current and futuristic capabilities:

1. Federated heterogeneous data architecture framework.
2. Service-oriented and layered architecture.
3. Application layers, offering domain-specific services and data portability.
4. Common services layer, such as data access, discovery, replica selection, task management, virtual data catalog, remote computation, remote visualization, and remote sensors.
5. Data systems software layer, with such information as metadata, formats, semantic standards, ontology, replica catalog, and security protocols.

Testbed   **Production**   *Interoperability Across Science Domains*

ESGF Data System Evolution

2012 – 2013	**2014 – 2015**	**2016**
<u>Foundation Development</u>	<u>Integration and Release</u>	<u>Evaluation and Deployment</u>

• **ESGF architecture refinement for climate project use case studies for diverse data sets**
   ○ **Provenance & Ontology**
• **ESGF collaborative distributed analysis infrastructure using UV-CDAT**
   ○ **Local and remote analysis**
   ○ **Enable reproducibility via workflow**
• **GIS integration**
• **Training and documentation**
	• **Expand to other climate projects & science domains**
	• **Full suite of server-side analysis and visualization**
	• **Machine learning for pattern discovery and prediction**
	• **Decision analytics based quantifying uncertainties**
	• **Streaming analysis, visualization and sensors**
	• **Model intercomparison metrics**
	• **Training and documentation**
		• **Evaluation and iterative science domain community feedback and upgrade**
		• **Debugging**
		• **Continued user feedback**
		• **Operational transition support by domain**
		• **Extended community training and documentation**

Climate   **ESGF Science Domains**   CMIP3, CMIP5, CESM, ARM, obs4MIPs, ana4MIPs, CORDEX, TAMIP, CDIAC, geoMIP, Ameriflux

Petabytes (10^15)   **Exabytes (10^18)**

Fig. 7.8 High-level roadmap for evolving ESGF across many climate science projects

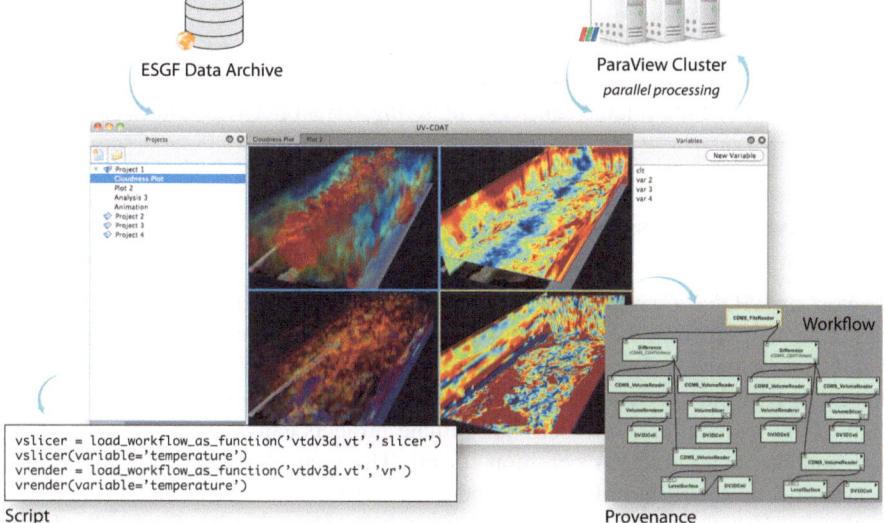

ESGF Data Archive   ParaView Cluster
                                               parallel processing

```
vslicer = load_workflow_as_function('vtdv3d.vt','slicer')
vslicer(variable='temperature')
vrender = load_workflow_as_function('vtdv3d.vt','vr')
vrender(variable='temperature')
```

Script   Provenance

Fig. 7.9 A glimpse of the ESGF infrastructure accessing distributed managed data at Leadership Computing Facilities. Parallel processing, performing data reduction and analysis, takes place via ParaView cluster analysis, and multiple views are displayed to the user. The workflow captures the entire process for reproducibility and knowledge sharing

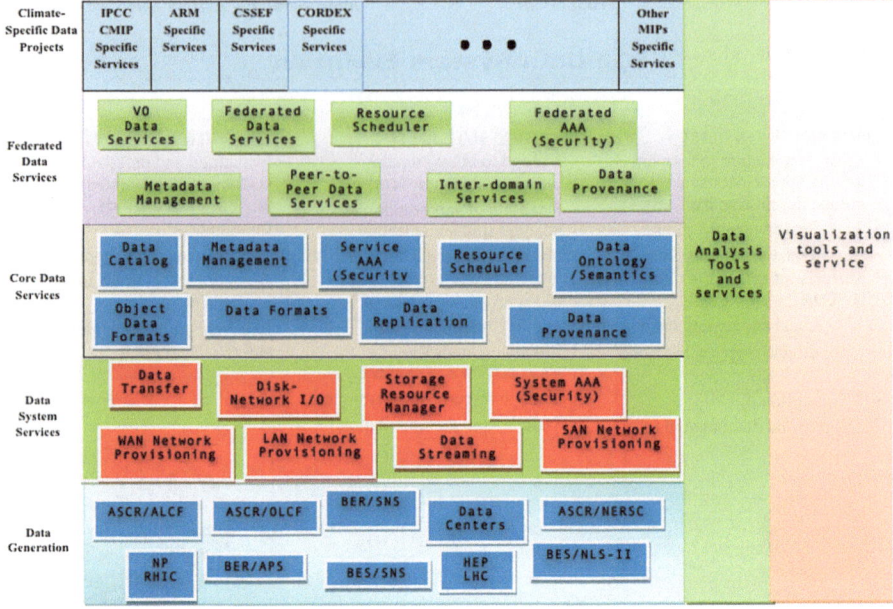

Fig. 7.10 Current and future end-to-end infrastructure for ESGF shows the framework and relationships for distributing climate science data and services

6. Data systems hardware, including storage systems, clusters, Leadership Computing Facilities, and display devices.
7. Networks and the related services, including virtual networks, network caches.

7.6 Outstanding Issues

As work on ESGF shows, building the infrastructure for extreme-scale computing and gathering support from the research community to sustain a distributed network are significant challenges. To continue to build on the successes of ESGF, we recommend that DOE BER host a data forum, where data systems and services architects from each of the national and international climate projects can discuss methodologies, philosophies, and standards common for all. The goal of such a forum would be to establish an open, common-component architecture for distributed science data systems and services within the greater community. However, the forum should not be limited to DOE, but open to the entire federation (including U.S. agencies such as the National Science Foundation, NASA, and NOAA and the international partners) to encourage the large-scale adoption of all approved standards—a long-term goal that DOE cannot accomplish in isolation. For this important national endeavor, which involves distributed data organization, archiving, and sharing of dispersed resources, we encourage partnering by all organizations.

7.7 Summary

ESGF is a key data-dissemination infrastructure and resource for climate simulation, observation, and reanalysis data. There are three major activities that affect the need for increased bandwidth: receiving data from all of the providers (including periodic updates), replicating that data to other national and international sites, and responding to requests from users for portions of the data holdings. Each activity requires network bandwidth the size of a petabyte data repository. For the future ESGF archives, the repository will increase by many orders of magnitude, but at best, network bandwidth will increase by only 1 order of magnitude in the next 5+ years. By replicating data sites, we can spread out the demand for services, which will resolve part of the gap. However, all ESGF sites need the fastest available network as soon as possible if the federation is to succeed at delivering results to prospective customers and large-scale data movement.

Acknowledgments Work described in this chapter performed under the auspices of the U.S. Department of Energy by Lawrence Livermore National Laboratory under Contract DE-AC52-07NA27344. The development and operation of ESGF and UV-CDAT is supported by the efforts of principal investigators, software engineers, data managers and system administrators from many agencies and institutions worldwide. ESGF contributors include ANL, ANU, BADC, CMCC, DKRZ, ESRL, GFDL, GSFC, JPL, IPSL, NCAR, LBNL, LLNL (leading institution), ORNL, PMEL, PNNL RPI, and SNL. ESGF funding provided by the U.S. Department of Energy, the U.S. National Atmospheric and Space Administration (NASA), and the European Infrastructure for the European Network for Earth System Modeling (IS-ENES). UV-CDAT contributors include GSFC, Knitwear, LANL, LBNL, LLNL (leading institution), NYC-Poly, ORNL, SCI and ESRL. UV-CDAT funding provided by U.S. Department of Energy, the U.S. National Atmospheric and Space Administration (NASA), and the U.S. National Oceanic and Atmospheric Administration (NOAA).

Chapter 8
Future Perspectives

Wolfgang Hiller and Reinhard Budich

Where will the development go? Whereas some skepticism still seems to be appropriate as to how far the initiatives and projects described above will have a lasting influence on the every-day-life of the single researcher, it seems to be out of question that even more and harder, joint efforts of the community are necessary to cope with the big data problem mentioned above.

At the lowest level, the speed with which model inter-comparison projects and distributed data processing can be executed very much depends upon the bandwidth available between cooperating research centers: The community needs to develop a much better leverage on organizations like Internet2 or GEANT to establish connectivity with much better performance between these centers worldwide. Software Defined Networking offering Quality of Service on demand might turn out to be helpful here.

Also on the infrastructure level, the security issue is of concern. The cumbersome process of granting access rights by generating community specific certificates for users of C3Grid or ESG is a working, but not a sustainable solution. These certificates are generally not approved by the certificate agencies worldwide, but serve as internal solution to grant access to scientists or research institutions belonging to the climate community: They cannot serve for interoperability between community Grids or Clouds worldwide. Clearly better solutions are needed. It has been demonstrated that an efficient and sustainable implementation of virtual organizations is hard to achieve, but should be kept high on the future agenda. Experiences from C3Grid and the connectivity to the ESG federation show that general solutions for an international

W. Hiller (✉)
Alfred-Wegener-Institut für Polar- und Meeresforschung, Am Handelshafen 12, 27570 Bremerhaven, Germany
e-mail: wolfgang.hiller@awi.de

R. Budich
Max-Planck-Institut für Meteorologie, Bundesstraße 53, 20146 Hamburg, Germany
e-mail: reinhard.budich@mpimet.mpg.de

W. Hiller et al., *Earth System Modelling – Volume 6*, SpringerBriefs in Earth System Sciences, DOI: 10.1007/978-3-642-37244-5_8, © The Author(s) 2013

interoperability are urgently needed. It remains to be seen if more recent initiatives and projects like IS-ENES and EUDAT can and will help in this respect.

Ease of use can also be fostered by easier access, easier in a legal sense. Initiatives worldwide have been requesting open access for data and publications which originate from publicly funded research. In Europe a recent initiative of the European Commission in the Seventh Framework Programme has put this higher on the political agenda. Some European governments have promised to grant free access to publicly funded research data in the next two years, either sectorial like in Germany for geospatial data or for the full scale of publicly funded research data and publications like in a recent initiative of the UK government. It remains to be seen what the influence of such initiaitves will be on the data access and distribution of climate research data, which have eminently high volumes and access problems on the technical side. Experience from the C3Grid community shows that it needs dedicated portals for the different stakeholders, instrumented with specific workflows to overcome the access and interpretation problems of climate research data, since "only" free access on its own will not be beneficial for the general public as potential user of climate research data.

Not only the workflows need to be optimized in such a way that every experiment writes out its own comprehensive set of metadata, also these metadata and their onthologies and vocabularies need constant development, maintenance and harmonization by all members of the communities involved, including an appropriate governance structure. Furthermore, provenance metadata will not be sufficient in the long run, content metadata need to be tackled one day due to the demands from the impacts side.

However the most important future perspective is the possibility to move over from the prototypical Grid-based workflows documented in this Volume to a sound distributed archive and processing basis for analysis workflows in the near future. This definitely can also help to address the imbalance problem of massive amounts of computing time dedicated to climate simulations but only a fraction of that used for thorough analysis of the massive amount of data generated: Post-processing and analysis of data in many cases need considerably more time than the production. This implies that the performance of the tools used and longevity of the data needed deserve more attention by the community.

A rather recent initiative tackles "Climate Knowledge Discovery" (CKD) and can be expected to amend current traditional workflows. Traditional methods probably cannot provide adequate tools for data intensive analysis needed to keep up with the growth of the data in climate science in all dimensions. To appropriately complement recent climate model simulations, CKD initiatives discuss the application of Complex Network algorithms to climate datasets and a combination with methods from the Semantic Web and knowledge discovery. For the future one could envisage that application of these techniques on a broader scale could provide unique insights into challenging features of the Earth system, including anomalies, nonlinear dynamics and risk analysis. The breakthroughs needed to address these challenges will only come from collaborative efforts like C3Grid, ESGF, IS-ENES, EUDAT and others who have paved the path for more comprehensive efforts in the future: These will involve several disciplines, including end-user scientists, specialists on large-scale graph analytics, semantic technologies and knowledge discovery algorithms as well as computer and computational scientists. Such a combination could finally be capable of doing research and implementation of knowledge discovery methods in climate science.

Glossary

AIRS Atmospheric Infrared Sounder

AJAX Asynchronous JavaScript and XML

AMIP Atmospheric Model Intercomparison Project

ANA4MIPs Reanalysis for the Coupled Model Intercomparison

ANL Argonne National Laboratory

*AR*4 Fourth Assessment Report

*AR*5 Fifth Assessment Report

ARMBE Atmospheric Radiation Measurement Best Estimate

ASCII American Standard Code for Information Interchange

BADC British Atmospheric Data Centre

BER Office of Biological and Environmental Research

C − LAMP Carbon Land Model Intercomparison Project

C3Grid Collaborative Climate Community Data and Processing Grid

CCSM Community Climate System Model

CDAT Climate Data Analysis Tools

CDIAC Carbon Dioxide Information Analysis Center

CEDA Centre of Environmental Data Archival

CERA Climate and Environmental Retrieval and Archive

CERES Clouds and Earth's Radiant Energy Systems

CESM Community Earth System Model

CF Climate and Forecast

CIM Common Information Model

CKD Climate Knowledge Discovery

*CMIP*5 Climate Modelling Intercomparison Project No. 5

CORDEX Coordinated Regional Climate Downscaling Experiment

COWS CEDA OGC Web Services

CSML Climate Science Modeling Language

CSS Climate Storage System

CSSEF Climate Science for a Sustainable Energy Future

CSW Catalogue Service specification for the Web

CWT Circulation Weather Types

DCMIP Dynamical Core Model Intercomparison Project

DDC Data Distribution Centre

DIS Data Information Service

DKRZ Deutsches Klimarechenzentrum—German Climate Computing Centre

DLESE Digital Library for Earth System Education

DOE U.S. Department of Energy

DOI Digital Object Identifier

DRS Date Reference Syntax

DTN Data Transfer Nodes

ESG Earth System Grid

ESGF Earth System Grid Federation

ESS Earth System Science

EUCLIPSE European Union Cloud Intercomparison, Process Study & Evaluation Project

EUDAT European Data Infrastructure

FTS Full-text search

GEANT GEometry ANd Tracking

GEO Group on Earth Observations

GeoMIP Geo-engineering Model Intercomparison Project

GEOSS Global Earth Observations System of Systems

GIS Geographical Information System

GML Geography Markup Language

GNDMS Generation N Data Management System

GridFTP Grid File Transfer Protocol

GSDI Global Spatial Data Infrastructures

GSI Grid Security Infrastructure

GTS Global Telecommunication System

GUI Graphical User Interface

IEEE Institute of Electrical and Electronics Engineers

INSPIRE Infrastructure for Spatial Information in Europe

IPCC Intergovernmental Panel on Climate Change

IS − ENES Infrastructure for the European Network for Earth System Modelling

jOAI Java-based open source Open Archives Initiative

JSDL Job Submission Description Language

KML Keyhole Markup Language

LANL Los Alamos National Laboratory

LAS Live Access Server

LBNL Lawrence Berkeley National Laboratory

LLNL Lawrence Livermore National Laboratory

LUCID Land-Use and Climate, Identification of Robust Impacts

MERRA Modern Era-Retrospective Analysis for Research and Applications

MERSEA Marine Environment and Security for the European Area

METAFOR Common Metadata for Climate Modelling Digital Repositories

MetOceanDWG Meteorology and Oceanography Domain Working Group

MISR Multi-angle Imaging SpectroRadiometer

MLS Microwave Limb Sounder

NARCCAP North American Regional Climate Change Assessment Program

NASA National Aeronautics and Space Administration

NCAR National Center for Atmospheric Research

NERC National Environment Research Council

NetCDF Network Common Data Form

NOAA National Oceanic and Atmospheric Administration

OAI Open Archives Initiative

OAI − PMH Open Archives Initiative Protocol for Metadata Harvesting

OASIS Organization for the Advancement of Structured Information Standards

obs4MIPs Observational Products More Accessible for Coupled Model Intercomparison

OECD Organisation for Economic Co-operation and Development

OGC Open Geospatial Consortium

OGF Open Grid Forum

OPeNDAP Open-source Project for a Network Data Access Protocol

OSG Open Science Grid

P2P Peer to peer

panFMP PANGAEA® Framework for Metadata Portals

PCMDI Program for Climate Model Diagnosis and Intercomparison

PMIP Paleoclimate Modeling Intercomparison Project

POP Parallel Ocean Program

QC Quality Control

RDF Resource Description Framework

REST Representational State Transfer

SAML Security Assertion Markup Language

SOAP Simple Object Access Protocol

SQL Structured Query Language

TAMIP Transpose-Atmospheric Model Intercomparison Project

TRMM Tropical Rainfall Measuring Mission

UI User Interface

UML Unified Modeling Language

URL Uniform Resource Locator

UV−CDAT Ultra-scale Visualization Climate Data Analysis Tools

W3C World Wide Web Consortium

WAN Wide-Area Network

WCS Web Coverage Service

WDCC World Data Center for Climate

WDC World Data Center

WDCC World Data Center for Climate

WMF Website Management Framework

WMS Web Map Service

WSDL Web Services Description Language

WSL Workflow Specification Language

WSRF Web Services Resource Framework

WSS Workflow Scheduling Service

XML Extensible Markup Language

Index